ABSTRACT ALGEBRA: A FIRST COURSE

DAN SARACINO

Colgate University

WAVELAND PRESS, INC.

Prospect Heights, Illinois

For information about this book, write or call:

Waveland Press, Inc.
P.O. Box 400
Prospect Heights, Illinois 60070
(708) 634-0081

PREFACE

This book is intended for use in a one-semester junior-senior level course in abstract algebra.

There is more material than can be covered in one term, so as to allow some flexibility in using the book. Sections 0–13 cover preliminary topics and the basics of group theory, and Sections 16–20 cover ring theory, up through quotient fields and the construction of field extensions as quotients of polynomial rings. The optional Sections 14 and 15, on finite abelian groups and Sylow theorems, allow extended coverage of groups. (These sections are arranged so that it is easy to cover the statements and applications of the results without getting into the proofs.) The optional Section 21 provides more exposure to rings.

One of my main goals in writing the book has been to include an unusually large number of examples, in order to help clarify the abstract concepts as they arise. Another goal has been to make the proofs of the theorems, especially the early ones on groups, do more than prove the stated results. I have attempted to convey some impression of where the proofs came from and why they proceed as they do. For many students, learning how to write proofs is the hardest part of abstract algebra; I have tried to write a book that will be helpful in overcoming this difficulty.

The order in which the various topics are presented is fairly standard, for a groups-before-rings approach. Specific decisions as to "what should come before what" were made with three objectives in mind:

1. To present as many examples as soon as possible;

2. To do easier things first;

3. To avoid placing a large block of background material (number theory, functions, equivalence relations) before the beginning sections on groups.

The exercises, for the most part, range from easy to moderately difficult. Most of them ask for understanding of the ideas involved, rather than flashes of insight. However, there are some challenging ones too, to keep things interesting.

Preliminary versions of this book have been class-tested at Colgate University over the last several years. I have tried to include comments about points that have caused students trouble in the past, in the hope of heading off problems in the future. I think my students have had a beneficial impact on the book, and I want to express my appreciation to them. Thanks are also due my colleagues, Dave Lantz and Chris Nevison, who used the manuscript as a text and offered helpful comments about it.

It is a pleasure for me to acknowledge the valuable assistance of the following reviewers: F. Doyle Alexander (Stephen F. Austin State University), Joseph B. Dennin, Jr. (Fairfield University), James E. Dowdy (West Virginia University), Mark P. Hale, Jr. (University of Florida), William F. Keigher (Rutgers University, Newark Campus), and William Trench (Drexel University). Last, but not least, I want to thank Lorraine Aveni for doing the bulk of the typing, always with her characteristic good humor and efficiency.

Hamilton, New York   D. S.
December 1979

CONTENTS

SETS AND INDUCTION

One of the most fundamental notions in any part of mathematics is that of a *set*. You are probably already familiar with the basics about sets, but we will start out by running through them quickly, if for no other reason than to establish some notational conventions. After these generalities, we will make some remarks about the set of positive integers, and in particular about the method of mathematical induction, which will be useful to us in later proofs.

For us, a **set** will be just a collection of entities, called the **elements** or **members** of the set. We indicate that some object x is an element of a set S by writing $x \in S$. If x is not an element of S, we write $x \notin S$.

In order to specify a set S, we must indicate which objects are elements of S. If S is finite, we can do this by writing down all the elements inside braces. For example, we write

$$S = \{1, 2, 3, 4\}$$

to signify that S consists of the positive integers 1, 2, 3, and 4. If S is infinite, then we cannot list all its elements, but sometimes we can give enough of them to make it clear what set S is. For instance,

$$S = \{1, 4, 7, 10, 13, 16, \dots\}$$

indicates the set of all positive integers that are of the form $1 + 3k$ for some nonnegative integer k.

We can also specify a set by giving a criterion that determines which objects are in the set. Using this method, the set $\{1, 2, 3, 4\}$ could be denoted by

$$\{x \mid x \text{ is a positive integer} \leqslant 4\},$$

where the vertical bar stands for the words "such that." Likewise, the set $\{1, 4, 7, 10, 13, 16, \dots\}$ could be written as

$$\{x \mid x = 1 + 3k \quad \text{for some nonnegative integer } k\}.$$

1

Some sets occur so frequently that it is worthwhile to adopt special notations for them. For example, we use

$\mathbb{Z}$ to denote the set of all integers,

$\mathbb{Q}$ to denote the set of all rational numbers,

$\mathbb{R}$ to denote the set of all real numbers, and

$\mathbb{C}$ to denote the set of all complex numbers.

The symbol $\varnothing$ denotes the **empty set** or **null set**, i.e., the set with no elements.

Sometimes we wish to express the fact that one set is included in another, i.e., that every element of the first set is also an element of the second set. We do so by saying that the first set is a *subset* of the second.

DEFINITION If S and T are sets, then we say that S is a **subset** of T, and write $S \subseteq T$, if every element of S is an element of T.

Examples If $S = \{1,2,3\}$ and $T = \{1,2,3,4,5\}$, then $S \subseteq T$.
   If $S = \{\pi, \sqrt{2}\}$ and $T = \{\pi, 5, \sqrt{2}\}$, then $S \subseteq T$. We write $S \not\subseteq \{5, \sqrt{2}\}$ because $\pi \in S$ but $\pi \notin \{5, \sqrt{2}\}$.
   If we let

$$\mathbb{Z}^+ = \{x \mid x \text{ is a positive integer}\},$$

then $\mathbb{Z}^+ \subseteq \mathbb{Z}$; similarly, we have $\mathbb{Q}^+ \subseteq \mathbb{Q}$ and $\mathbb{R}^+ \subseteq \mathbb{R}$.

Observe that for any set S, $S \subseteq S$, that is, S is a subset of itself. Also observe that $\varnothing \subseteq S$, no matter what set S is. Perhaps the best way to see this is to ask yourself how it could be false. If $\varnothing \not\subseteq S$, then there is some $x \in \varnothing$ which is not in S; but this is nonsense, because there is no $x \in \varnothing$, period.

We say that two sets S and T are **equal**, and we write $S = T$, if S and T have the same elements. Clearly, then, saying that $S = T$ is equivalent to saying that both $S \subseteq T$ and $T \subseteq S$. If $S \subseteq T$ but $S \neq T$, we say that S is a **proper** subset of T. If we wish to emphasize that S is a proper subset, we write $S \subsetneq T$.

Very often we consider sets that are obtained by performing some operation on one or more given sets. For example, if S and T are sets, then their **intersection**, denoted by $S \cap T$, is defined by

$$S \cap T = \{x \mid x \in S \text{ and } x \in T\}.$$

The **union** of S and T, denoted by $S \cup T$, is given by

$$S \cup T = \{x \mid x \in S \text{ or } x \in T \text{ or both}\}.$$

The union and intersection of more than two sets are defined in an analogous way; for instance,

$$S \cap T \cap U = \{x \mid x \in S \text{ and } x \in T \text{ and } x \in U\}.$$

Examples Let $S = \{1,2,3,4,5\}$ and $T = \{2,4,6\}$. Then $S \cap T = \{2,4\}$ and $S \cup T = \{1,2,3,4,5,6\}$.
Again let $S = \{1,2,3,4,5\}$. Then $S \cap \mathbb{R} = S$ and $S \cup \mathbb{R} = \mathbb{R}$.

We can illustrate many of the notions we have introduced by generalizing this last example.

THEOREM 0.1 Let S and T be sets. Then $S \subseteq T$ if and only if $S \cap T = S$.

PROOF. We must show that $S \subseteq T$ implies $S \cap T = S$, and that conversely $S \cap T = S$ implies $S \subseteq T$.
    Assume $S \subseteq T$. To show that $S \cap T = S$ we have to show that $S \cap T \subseteq S$ and $S \subseteq S \cap T$. The first is clear; for the second we must show that every element of S is an element of S and of T. Clearly any element of S is an element of S; and since we are assuming $S \subseteq T$, any element of S is an element of T, too, so we are done with the first half of the proof.
    Now assume $S \cap T = S$; we show that $S \subseteq T$. Why is it true that any element of S is an element of T? Because any element of S is an element of $S \cap T$ by our assumption, and any element of $S \cap T$ is clearly an element of T. $\square$

It is also true that $S \subseteq T$ if and only if $S \cup T = T$. The proof of this is left as an exercise.
    As a matter of notation, we adopt the abbreviation "iff" for "if and only if." Thus we say that $S \subseteq T$ iff $S \cup T = T$. Sometimes the symbol $\Leftrightarrow$ is used in place of iff; using $\Leftrightarrow$, we would write $S \subseteq T \Leftrightarrow S \cup T = T$.
    One set that is particularly important in mathematics is the set $\mathbb{Z}^+$ of positive integers. We will see that, in abstract algebra, concepts defined in terms of positive integers can often help to clarify what is going on. For this reason, methods for working with integers can be very valuable tools. Perhaps the most useful strategy for proving things about $\mathbb{Z}^+$ is the method of **mathematical induction**.
    Suppose we have in mind a statement $P(n)$ about the integer n. For example, $P(n)$ might say "n is even," or "n is the square of some integer," or "If p is a prime, then every group of order p^n has nontrivial center" (whatever that means). Mathematical induction provides us with a way of trying to prove that $P(n)$ is true for every positive n.

The technique rests on an intuitively acceptable axiom called the

Well-Ordering Principle: Every nonempty subset of $\mathbb{Z}^+$ has
a smallest element.

The well-ordering principle yields two slightly different forms of induction, both of which are good to know.

THEOREM 0.2 (Mathematical Induction, first form) Suppose $P(n)$ is a statement about positive integers, and we know two things:

i) $P(1)$ is true;

ii) for every positive m, if $P(m)$ is true, *then* $P(m+1)$ is true.

Under these circumstances, we can conclude that $P(n)$ is true for all positive n.

PROOF. Suppose $P(n)$ is false for some positive n. Then $S = \{n \mid n \in \mathbb{Z}^+$ and $P(n)$ is false$\}$ is a nonempty subset of $\mathbb{Z}^+$. By the well-ordering principle, S has a smallest element, say n_0. Clearly $n_0 \neq 1$, because $P(1)$ is true by (i). Therefore, $n_0 - 1$ is a positive integer, and $P(n_0 - 1)$ is true because $n_0 - 1$ is smaller than n_0. By (ii), this means that $P(n_0 - 1 + 1)$ is true; that is, $P(n_0)$ is true, and this contradicts the fact that $P(n_0)$ is false!

Since the supposition that $P(n)$ is false for some n has led us to a contradiction, we conclude that $P(n)$ holds for all $n \in \mathbb{Z}^+$. $\square$

What you do to prove something by induction, then, is this. You first show that $P(1)$ is true (this is usually trivial). You then show that for an arbitrary positive m, if $P(m)$ is true, *then* $P(m+1)$ is true. You do this by assuming that $P(m)$ is true and using this assumption to establish that $P(m+1)$ is true.

Sometimes people are bothered by the word "assuming" in the last sentence. They get worried that "assuming that $P(m)$ is true" amounts to assuming what we are trying to prove. *But it does not*, for we are not assuming that $P(m)$ is true for all m. Rather we are arguing, for an arbitrary fixed m, that if $P(m)$ is true for that m, then so is $P(m+1)$. The only way of doing this is to show that $P(m+1)$ is true on the basis of the assumption that $P(m)$ is.

Examples

1. You may recall that, in calculus, when you are evaluating definite integrals from the definition as a limit of Riemann sums, you run into sums such as $1 + 2 + 3 + \cdots + n$, and you need formulas for these sums in terms of n. The formula for $1 + 2 + \cdots + n$, for example, is $n(n+1)/2$. Let's prove it, by induction.

We take for $P(n)$ the statement that

$$1 + 2 + \cdots + n = n(n+1)/2;$$

we hope to show that $P(n)$ is true for all positive n. First we check $P(1)$: It says that $1 = 1(1 + 1)/2$, which is certainly true. Second, we assume that $P(m)$ is true for some arbitrary positive m, and we use this assumption to show that $P(m + 1)$ is true. That is, we assume that

$$1 + 2 + \cdots + m = m(m + 1)/2 \qquad \text{[0.1]}$$

and we try to show that

$$1 + 2 + \cdots + m + m + 1 = (m + 1)(m + 1 + 1)/2. \qquad \text{[0.2]}$$

The natural thing to do is to add $m + 1$ to both sides of Eq. [0.1]:

$$1 + 2 + \cdots + m + m + 1 = [m(m + 1)/2] + m + 1.$$

Now

$$[m(m + 1)/2] + m + 1 = [m(m + 1) + 2(m + 1)]/2 = [(m + 1)(m + 2)]/2,$$

so we have Eq. [0.2].

Therefore, by induction, $P(n)$ holds for all positive n, and we are done.

2. A similar formula, also used in calculus, is

$$1^2 + 2^2 + \cdots + n^2 = \frac{n(n + 1)(2n + 1)}{6}.$$

If we take this equation as $P(n)$, then $P(1)$ says

$$1^2 = \frac{1(1 + 1)(2 + 1)}{6},$$

which is true. To show that $P(m)$ implies $P(m + 1)$, we assume that

$$1^2 + 2^2 + \cdots + m^2 = \frac{m(m + 1)(2m + 1)}{6}, \qquad \text{[0.3]}$$

and we try to show that

$$1^2 + 2^2 + \cdots + (m + 1)^2 = \frac{(m + 1)(m + 2)[2(m + 1) + 1]}{6}. \qquad \text{[0.4]}$$

If we add $(m + 1)^2$ to both sides of Eq. [0.3], we conclude that

$$1^2 + 2^2 + \cdots + m^2 + (m + 1)^2 = \frac{m(m + 1)(2m + 1)}{6} + (m + 1)^2,$$

and the right-hand side is

$$\frac{m(m + 1)(2m + 1) + 6(m + 1)^2}{6} = \frac{(m + 1)[m(2m + 1) + 6(m + 1)]}{6}$$

$$= \frac{(m + 1)(2m^2 + 7m + 6)}{6} = \frac{(m + 1)(m + 2)(2m + 3)}{6},$$

so we have Eq. [0.4].

Thus our formula is established by induction.

3. The following popular example illustrates the fact that some care is necessary in trying to prove something by induction. We shall "prove" that all horses are the same color.

Let $P(n)$ be the statement: "For every set of n horses, all the horses in the set are the same color." We "prove" $P(n)$ for all n by induction. Clearly, $P(1)$ is true since any horse is the same color as itself. Now assume that $P(m)$ is true and let us show that $P(m+1)$ holds. Let S be a set of $m+1$ horses; say the horses in S are $h_1, h_2, \ldots, h_{m+1}$ (h for horse). Now $h_1, h_2, \ldots, h_m$ comprise a set of m horses, so since $P(m)$ holds, $h_1, h_2, \ldots, h_m$ are all the same color. Likewise, $h_2, h_3, \ldots, h_{m+1}$ make up a set of m horses, so $h_2, h_3, \ldots, h_{m+1}$ are all the same color. Combining these statements, we see that all $m+1$ horses are the same color (for instance, they are all the same color as h_2).

There must be something wrong with this, but what?

The second form of induction is similar to the first, except that (ii) is modified somewhat.

THEOREM 0.3 (Mathematical Induction, second form) Suppose $P(n)$ is a statement about positive integers and we know two things:

i) $P(1)$ is true;

ii) for every positive m, *if $P(k)$ is true for all positive $k < m$, then $P(m)$ is* true.

Under these circumstances, we can conclude that $P(n)$ is true for all positive n.

The verification that this works is like that for the first form, and we leave it as an exercise.

The part that one assumes in trying to establish (ii) in a proof by induction (either form) is called the **inductive hypothesis**. Thus in the first form, the inductive hypothesis is the assumption that $P(m)$ is true, from which we try to argue that $P(m+1)$ is true. In the second form, the inductive hypothesis is that $P(k)$ is true for all positive $k < m$, and our task is to use this to show that $P(m)$ holds.

Which form of induction one should use in any particular case depends on the situation at hand. We have seen some examples where the first form is appropriate, and we would like to close with a significant instance where the second form is called for.

If a and b are integers, then we say that a **divides** b, and we write $a|b$, if there is an integer c such that $b = ac$. If no such c exists, we write $a \nmid b$. For instance, $2|6$, $7|14$, and $3 \nmid 10$. A positive integer p is called a prime if $p > 1$ and the only positive integers that divide p are 1 and p. The first few primes are 2, 3, 5, 7, 11, 13, 17, ...

The *Fundamental Theorem of Arithmetic* asserts that every positive $n > 1$ can be written as the product of finitely many primes, and that except for a possible rearrangement of the factors, there is only one such factorization.

THEOREM 0.4 (Fundamental Theorem of Arithmetic) Let $n > 1$. Then there are primes $p_1, p_2, p_3, \ldots, p_r$ (not necessarily distinct) such that $n = p_1 p_2 \cdots p_r$. Moreover, if $n = q_1 q_2 \cdots q_s$ is another such factorization, then $r = s$ and the p_i's are the q_j's, possibly rearranged.

PROOF. We prove only the existence of a factorization here. The proof of the uniqueness assertion requires some more groundwork and is left to the exercises in Section 4.

We wish to prove that $P(n)$ holds for every $n \geqslant 2$, where $P(n)$ says that n can be written as a product of primes. Accordingly, we start our induction at 2 rather than at 1; we show (see Exercise 0.13) that $P(2)$ is true and that for any $m > 2$, if $P(k)$ holds for all $2 \leqslant k < m$, then $P(m)$ holds. Clearly $P(2)$ is true, since 2 is itself a prime. Now take $m > 2$. If m is a prime, then $P(m)$ holds. If m is not a prime, then we can write $m = ab$, where neither a nor b is m. Thus $2 \leqslant a, b < m$, and by the inductive hypothesis we can write

$$a = p_1 p_2 \cdots p_t, \quad b = q_1 q_2 \cdots q_u$$

for primes p_i, q_j. Then

$$m = ab = p_1 p_2 \cdots p_t q_1 q_2 \cdots q_u,$$

and we have shown that m can be factored into primes. $\square$

Notice that all we can say about a and b here is that they are less than m, so we need the inductive hypothesis as in the second form of induction, in order for this proof to go through.

EXERCISES
Sets

0.1 Let $S = \{2, 5, \sqrt{2}, 25, \pi, 5/2\}$ and let $T = \{4, 25, \sqrt{2}, 6, 3/2\}$. Find $S \cap T$ and $S \cup T$.

0.2 With S and T as in Exercise 0.1, show that
$$\mathbb{Z} \cap (S \cup T) = (\mathbb{Z} \cap S) \cup (\mathbb{Z} \cap T) \quad \text{and} \quad \mathbb{Z} \cup (S \cap T) = (\mathbb{Z} \cup S) \cap (\mathbb{Z} \cup T).$$

0.3 Let S and T be sets. Prove that
$$S \cap (S \cup T) = S \quad \text{and} \quad S \cup (S \cap T) = S.$$

0.4 Let S and T be sets. Prove that $S \cup T = T$ iff $S \subseteq T$.

0.5 Let A, B, and C be sets. Prove that $A \cap (B \cup C) = (A \cap B) \cup (A \cap C)$.

0.6 Let A, B, and C be sets. Prove that $A \cup (B \cap C) = (A \cup B) \cap (A \cup C)$.

Induction

0.7 What's wrong with the proof that all horses are the same color?

0.8 Prove that

$$1^3 + 2^3 + 3^3 + \cdots + n^3 = \left[\frac{n(n+1)}{2} \right]^2$$

for all $n \geqslant 1$.

0.9 Prove that

$$1 + 3 + 5 + \cdots + (2n+1) = (n+1)^2$$

for all $n \geqslant 1$.

0.10 Prove that

$$2 + 4 + 6 + \cdots + 2n = n(n+1)$$

for all $n \geqslant 1$.

0.11 Prove Theorem 0.3.

0.12 Prove the following more general form of Theorem 0.2:

THEOREM. Suppose $P(n)$ is a statement about positive integers and c is some fixed positive integer. Assume

i) $P(c)$ is true; and

ii) for every $m \geqslant c$, if $P(m)$ is true, *then* $P(m+1)$ is true.

Then $P(n)$ is true for all $n \geqslant c$.

0.13 Prove the following more general version of Theorem 0.3:

THEOREM. Suppose $P(n)$ is a statement about positive integers and c is some fixed positive integer. Assume

i) $P(c)$ is true; and

ii) for every $m > c$, if $P(k)$ is true for all k such that $c \leqslant k < m$, *then* $P(m)$ is true.

Then $P(n)$ is true for all $n \geqslant c$.

0.14 Prove that

$$1 \cdot 2 + 2 \cdot 3 + 3 \cdot 4 + \cdots + (n-1)n = \frac{(n-1)(n)(n+1)}{3}$$

for all $n \geqslant 2$.

0.15 By trying a few cases, guess at a formula for

$$\frac{1}{1 \cdot 2} + \frac{1}{2 \cdot 3} + \frac{1}{3 \cdot 4} + \cdots + \frac{1}{(n-1)n}, \quad n \geqslant 2.$$

Try to prove that your guess is correct.

0.16 Prove that for every $n \geqslant 1$, 3 divides $n^3 - n$.

0.17 Prove that if a set S has n elements (where $n \in \mathbb{Z}^+$), then S has 2^n subsets.

0.18 The *Fibonacci sequence* $f_1, f_2, f_3, \ldots$ is defined as follows:

$$f_1 = f_2 = 1, \quad f_3 = 2, \quad f_4 = 3, \quad f_5 = 5, \quad f_6 = 8, \ldots,$$

and in general,

$$f_n = f_{n-1} + f_{n-2} \quad \text{for all } n \geqslant 3.$$

Prove that f_{5k} is divisible by 5 for every $k \geqslant 1$, that is, 5 divides every 5th member of the sequence.

BINARY OPERATIONS

In high school algebra, we spend a great deal of time solving equations involving real or complex numbers. At the heart of it, what we are really doing is answering questions about the addition and multiplication of these numbers.

In abstract algebra, we take a more general view, starting from the observation that addition and multiplication are both just ways of taking two elements and producing a third, in such a way that certain laws are obeyed. We study the situation where we have a set and one or more "operations" for producing outputs from given inputs, subject to some specified rules.

From this description, it is probably not clear to you why abstract algebra is any more profitable than going off and counting the grains of sand on the nearest beach. But it can be very profitable, for a number of reasons. For one thing, the abstract approach may clarify our thinking about familiar situations by stripping away irrelevant aspects of what is happening. For another, it may lead us to consider new systems that are valuable because they shed light on old problems. Yet again, a general approach can save us effort by dealing with a number of specific situations all at once. And finally, although it can take some time to appreciate this, abstraction can be just plain beautiful.

DEFINITION If S is a set, then a **binary operation** * on S is a function that associates to each ordered pair (s_1, s_2) of elements of S an element of S, which we denote by $s_1 * s_2$.

Observe that the definition says *ordered* pair. Thus (s_1, s_2) is not necessarily the same thing as (s_2, s_1), and $s_1 * s_2$ is not necessarily the same thing as $s_2 * s_1$. Notice too that * must assign an element of S to each and every pair (s_1, s_2), including those pairs in which s_1 and s_2 are the same element of S.

Examples

1. Addition is a binary operation on $\mathbb{Z}^+ : a * b = a + b$. Subtraction $(a * b = a - b)$ is not; but subtraction is a binary operation on $\mathbb{Z}$.

2. Multiplication $(a * b = a \cdot b)$ is a binary operation on $\mathbb{Z}^+$ or $\mathbb{Z}$ or $\mathbb{R}$ or $\mathbb{Q}$. Division $(a * b = a / b)$ is a binary operation on $\mathbb{Q}^+$ or $\mathbb{R}^+$ but not $\mathbb{Z}$ or $\mathbb{Z}^+$ or $\mathbb{R}$ or $\mathbb{Q}$.

3. $a * b = a^3 + b^2 + 1$ is a binary operation on $\mathbb{Z}$, $\mathbb{Q}$, $\mathbb{R}$, $\mathbb{Z}^+$, $\mathbb{Q}^+$, or $\mathbb{R}^+$.

4. Let X be some set and let S be the set of all subsets of X. For example, if $X = \{1\}$, then $S = \{\varnothing, \{1\}\}$, and if $X = \{1,2\}$, then

$$S = \{\varnothing, \{1\}, \{2\}, \{1,2\}\}.$$

The operation of intersection is a binary operation on S, since if A, B are elements of S, then $A * B = A \cap B$ is an element of S. Similarly, union gives us another binary operation on S.

5. Let S be the set of all 2×2 matrices with real entries. Thus an element of S looks like $\begin{pmatrix} a & b \\ c & d \end{pmatrix}$, where $a, b, c, d \in \mathbb{R}$. Define $x * y$ to be the matrix product of x and y, that is,

$$\begin{pmatrix} a & b \\ c & d \end{pmatrix} * \begin{pmatrix} e & f \\ g & h \end{pmatrix} = \begin{pmatrix} ae + bg & af + bh \\ ce + dg & cf + dh \end{pmatrix}.$$

Then $*$ is a binary operation on S. For instance,

$$\begin{pmatrix} 1 & 2 \\ 0 & 1 \end{pmatrix} * \begin{pmatrix} 2 & 1 \\ 1 & 2 \end{pmatrix} = \begin{pmatrix} 4 & 5 \\ 1 & 2 \end{pmatrix} \quad \text{and} \quad \begin{pmatrix} 3 & \pi \\ 1 & 1 \end{pmatrix} * \begin{pmatrix} 0 & 1 \\ 1 & -1 \end{pmatrix} = \begin{pmatrix} \pi & 3 - \pi \\ 1 & 0 \end{pmatrix}.$$

The definition of *binary operation* doesn't impose any restrictions on $*$, and in general a binary operation can be wildly misbehaved. Ordinarily, we want to consider operations that have at least something in common with familiar, concrete examples.

DEFINITIONS If $*$ is a binary operation on S, then $*$ is called **commutative** if $s_1 * s_2 = s_2 * s_1$ for every $s_1, s_2 \in S$. On the other hand, $*$ is called **associative** if $(s_1 * s_2) * s_3 = s_1 * (s_2 * s_3)$ for every $s_1, s_2, s_3 \in S$.

Examples

1. Subtraction on $\mathbb{Z}$ is neither commutative nor associative. For example,

$$1 - 2 \neq 2 - 1 \quad \text{and} \quad (1 - 2) - 3 \neq 1 - (2 - 3).$$

2. Let $a * b = 2(a + b)$ on $\mathbb{Z}$; then clearly $*$ is commutative. Now

$$(a * b) * c = 2(a + b) * c = 2(2(a + b) + c) = 4a + 4b + 2c,$$

and

$$a*(b*c)=a*2(b+c)=2(a+2(b+c))=2a+4b+4c.$$

Since these are not always the same, $*$ is not associative.

3. Let $a*b=2ab$ on $\mathbb{Z}$. Commutativity is again clear, and since

$$(a*b)*c=2ab*c=4abc \quad \text{and} \quad a*(b*c)=a*2bc=4abc,$$

$*$ is also associative in this case.

4. Multiplication of 2×2 matrices is associative:

$$\left(\begin{pmatrix} a & b \\ c & d \end{pmatrix}*\begin{pmatrix} e & f \\ g & h \end{pmatrix}\right)*\begin{pmatrix} i & j \\ k & l \end{pmatrix}=\begin{pmatrix} a & b \\ c & d \end{pmatrix}*\left(\begin{pmatrix} e & f \\ g & h \end{pmatrix}*\begin{pmatrix} i & j \\ k & l \end{pmatrix}\right),$$

as may be verified by multiplying out both sides. This operation is not commutative, however. For example,

$$\begin{pmatrix} 0 & 1 \\ 1 & 0 \end{pmatrix}*\begin{pmatrix} 1 & 2 \\ 3 & 4 \end{pmatrix}=\begin{pmatrix} 3 & 4 \\ 1 & 2 \end{pmatrix} \quad \text{but} \quad \begin{pmatrix} 1 & 2 \\ 3 & 4 \end{pmatrix}*\begin{pmatrix} 0 & 1 \\ 1 & 0 \end{pmatrix}=\begin{pmatrix} 2 & 1 \\ 4 & 3 \end{pmatrix}.$$

Before we consider a slightly more complicated example, observe that these first four examples say something about the relationship between commutativity and associativity—namely, that there isn't any! A binary operation can be commutative with or without being associative, and it can be noncommutative with or without being associative.

5. We introduce another operation on sets: If A,B are sets, then $A \triangle B$ denotes the **symmetric difference** of A and B. This is by definition the set of all elements that belong to either A or B, but not to both. Thus

$$A \triangle B=(A-B)\cup(B-A),$$

where $A-B$ denotes the set of elements of A that are not elements of B, and $B-A$ denotes the set of elements of B that are not elements of A. We can also write

$$A \triangle B=(A\cup B)-(A\cap B).$$

In Fig. 1.1, if A is the set of points inside the square, and B is the set of points inside the circle, then $A \triangle B$ consists of the points in the shaded region:

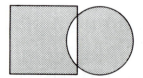

Figure 1.1

For example, if $A = \{1,2,3,4,5\}$ and $B = \{2,4,6\}$, then $A \triangle B = \{1,3,5,6\}$. If

$$A = \{x \mid x \in \mathbb{Z} \text{ and } x \geqslant 0\} \quad \text{and} \quad B = \{x \mid x \in \mathbb{Z} \text{ and } x \leqslant 0\},$$

then

$$A \triangle B = \{x \mid x \in \mathbb{Z} \text{ and } x \neq 0\}.$$

For any set A, $A \triangle A = \varnothing$ and $A \triangle \varnothing = A$.

Now let X be a set and let S be the set of subsets of X. Then $A * B = A \triangle B$ defines a binary operation on S. It is easy to see that $\triangle$ is commutative (Exercise 1.7). We claim that $\triangle$ is an associative operation on S; that is, if A, B, C are subsets of X, then

$$(A \triangle B) \triangle C = A \triangle (B \triangle C).$$

To verify this we have to show that each side of the equation is contained in the other. Let's assume that $x \in (A \triangle B) \triangle C$ and show that $x \in A \triangle (B \triangle C)$.

Since $x \in (A \triangle B) \triangle C$, we have either:

Case I. $x \in (A \triangle B)$ and $x \notin C$,   or

Case II. $x \notin (A \triangle B)$ and $x \in C$.

In Case I, $x \in (A \triangle B)$ implies that either $x \in A$ and $x \notin B$ or $x \notin A$ and $x \in B$. If the first of these holds, then we have $x \in A, x \notin B, x \notin C$, so that $x \in A$ and $x \notin (B \triangle C)$, so $x \in A \triangle (B \triangle C)$. If the second holds, then we have $x \notin A, x \in B, x \notin C$, so that $x \notin A$ and $x \in (B \triangle C)$, so $x \in A \triangle (B \triangle C)$.

In Case II, $x \notin A \triangle B$ implies that either $x \in A$ and $x \in B$, or $x \notin A$ and $x \notin B$. If the first alternative holds, then we have $x \in A$, $x \in B$, and $x \in C$, so that $x \in A$ and $x \notin (B \triangle C)$, and therefore $x \in A \triangle (B \triangle C)$. If the second alternative holds, then we have $x \notin A$, $x \notin B$, and $x \in C$, so $x \notin A$ and $x \in (B \triangle C)$, and therefore $x \in A \triangle (B \triangle C)$.

This completes the proof that $(A \triangle B) \triangle C \subseteq A \triangle (B \triangle C)$. The proof of the reverse inclusion is left as an exercise.

Before leaving these ideas it might be well to say a bit more about the need for talking about them in the first place. We are all so familiar with the commutativity and associativity of addition and multiplication on the integers, rationals, and reals that we take them for granted. It never occurs to us that they could be called into question. The point is that as we broaden our perspective and consider less familiar systems, we run into many where commutativity and associativity no longer hold true. Thus we have to be careful in working with such systems; we can only use commutativity and associativity in carrying out calculations if we have first ascertained that they happen to hold for the system we are working with. Calculations with the integers are easy because we can replace xy by yx whenever we feel like it, and we never have to worry about $(x+y)+z$ being the same as $x+(y+z)$. In general, we are not so lucky, and we do have to worry about such things.

EXERCISES

1.1 For the given sets S and T, find $S \triangle T$.

a) $S = \{2,5,\sqrt{2},25,\pi,5/2\}$, $T = \{4,25,\sqrt{2},6,3/2\}$

b) $S = \left\{ \begin{pmatrix} 1 & 3 \\ 4 & 6 \end{pmatrix}, \begin{pmatrix} 2 & 1 \\ 1 & 2 \end{pmatrix}, \begin{pmatrix} 5 & 8 \\ 0 & -1 \end{pmatrix}, \begin{pmatrix} 1 & 1 \\ 1 & \pi \end{pmatrix} \right\}$,

$T = \left\{ \begin{pmatrix} 2 & 1 \\ 1 & 2 \end{pmatrix}, \begin{pmatrix} 5 & 8 \\ 0 & 1 \end{pmatrix}, \begin{pmatrix} 1 & 2 \\ 3 & 4 \end{pmatrix} \right\}$

1.2 If A, B, and C are the sets of points inside the three circles below, what region represents $(A \triangle B) \triangle C$?

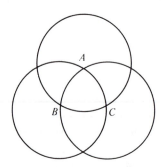

1.3 In each case, determine whether or not the given $*$ is a binary operation on the given set S.

a) $S = \mathbf{Z}$, $a*b = a+b^2$

b) $S = \mathbf{Z}$, $a*b = a^2 b^3$

c) $S = \mathbf{R}$, $a*b = \dfrac{a}{a^2+b^2}$

d) $S = \mathbf{Z}$, $a*b = \dfrac{a^2+2ab+b^2}{a+b}$

e) $S = \mathbf{Z}$, $a*b = a+b-ab$

f) $S = \mathbf{R}$, $a*b = b$

g) $S = \{1,-2,3,2,-4\}$, $a*b = |b|$

h) $S = \{1,6,3,2,18\}$, $a*b = ab$.

i) $S =$ the set of all 2×2 matrices with real entries, and if

$$a = \begin{pmatrix} r_1 & r_2 \\ r_3 & r_4 \end{pmatrix} \quad \text{and} \quad b = \begin{pmatrix} r_5 & r_6 \\ r_7 & r_8 \end{pmatrix},$$

then

$$a*b = \begin{pmatrix} r_1+r_5 & r_2+r_6 \\ r_3+r_7 & r_4+r_8 \end{pmatrix}$$

j) $S =$ the set of all subsets of a set X, $A*B = (A \triangle B) \triangle B$.

1.4 Is division a commutative operation on $\mathbb{R}^+$? Is it associative?

1.5 If S is the set of subsets of a set X, is the operation of intersection commutative on S? Is it associative?

1.6 For each case in Exercise 1.3 in which $*$ is a binary operation on S, determine whether $*$ is commutative and whether it is associative.

1.7 Show that symmetric difference is a commutative operation.

1.8 Complete the proof that symmetric difference is an associative operation.

1.9 If S is a finite set, then we can define a binary operation on S by writing down all the values of $s_1 * s_2$ in a table. For instance, if $S = \{a,b,c,d\}$, then the following gives a binary operation on S.

Columns

	$*$	a	b	c	d
	a	a	c	b	d
Rows	b	c	a	d	b
	c	b	d	a	c
	d	d	b	c	a

Here, for $s_1, s_2 \in S$, $s_1 * s_2$ is the element in row s_1 and column s_2. For example, $c * b = d$.

Is the above binary operation commutative? Is it associative?

GROUPS

As we have said, algebra is the study of various kinds of abstract systems. One of the most fundamental of these systems is the kind known as a *group*. In this section we shall get acquainted with this concept.

Certainly one of the attractive features of modern mathematics is the intertwining that takes place among its different branches. If you study algebra, analysis, topology, or mathematical logic, for example, you will see that certain ideas recur in all of them. The notion of a group is one of these ideas that seem to crop up everywhere. Beyond that, groups are important in areas where mathematics is used as a tool, such as chemistry, quantum mechanics, and elementary particle physics.

DEFINITION Suppose that:

i) G is a set and $*$ is a binary operation on G,

ii) $*$ is associative,

iii) there is an element e in G such that $x * e = e * x = x$ for all x in G, and

iv) for each element $x \in G$ there is an element $y \in G$ such that $x * y = y * x = e$.

Then G, together with the binary operation $*$, is called a **group**, and is denoted by $(G, *)$.

Observe that (iii) implies that G is nonempty. The element e in (iii) is called an **identity** element of G (we shall soon see that there is only one such element, so we will be able to call it *the* identity element of G). The element y in (iv) is called an **inverse** of x; we shall see that any element x has only one inverse, so we can call it *the* inverse of x. It is worth emphasizing the fact that the single identity element e satisfies $x * e = e * x = x$ for *any* choice of x, while the y in (iv) depends on x. We shall see that two distinct elements of G can never have the same inverse, so that if we picked a different x we would have a different y.

It is also worth emphasizing that we do *not* assume that $*$ is commutative. Groups with the special added property that their operation *is* commutative are called **abelian**, after the Norwegian mathematician Niels Abel (1802–1829).

Before we begin investigating the properties of groups in general, we will examine a number of examples in order to try to convey some impression of the universality of this notion. Historically, of course, the examples preceded the abstract definition; the abstract concept of "group" came into being only when people realized that many objects they were working with had various structural characteristics in common, and got the idea that some economy of effort could be achieved by studying these common features abstractly rather than over and over again in each separate case. In discussing the development of the subject, E. T. Bell has remarked that "wherever groups disclosed themselves, or could be introduced, simplicity crystallized out of comparative chaos."

We begin with the most familiar examples.

1. $(\mathbb{Z}, +)$. 0 is an identity element, and $-x$ is an inverse for x:

$$x + (-x) = (-x) + x = 0.$$

We are all familiar with the fact that $+$ is an associative binary operation on $\mathbb{Z}$.

Similarly, $(\mathbb{Q}, +)$ and $(\mathbb{R}, +)$ are groups.

2. $(\mathbb{Q}^+, \cdot)$. Multiplication is an associative binary operation on $\mathbb{Q}^+$, 1 is an identity element, and $1/x$ is an inverse for x since $x \cdot (1/x) = (1/x) \cdot x = 1$.

Similarly, $(\mathbb{R}^+, \cdot)$ is a group. Observe, however, that $(\mathbb{Z}^+, \cdot)$ is *not* a group, since no element other than 1 has an inverse. $(\mathbb{R} - \{0\}, \cdot)$ and $(\mathbb{Q} - \{0\}, \cdot)$ are groups. The set $\mathbb{R}^-$ of negative real numbers under multiplication is not a group because multiplication is not a binary operation on $\mathbb{R}^-$.

3. The set of n-tuples $(a_1, \ldots, a_n)$ of real numbers is a group under the binary operation given by addition of n-tuples:

$$(a_1, a_2, \ldots, a_n) * (b_1, b_2, \ldots, b_n) = (a_1 + b_1, a_2 + b_2, \ldots, a_n + b_n).$$

$(0, 0, \ldots, 0)$ is an identity element and $(-a_1, -a_2, \ldots, -a_n)$ is an inverse for $(a_1, a_2, \ldots, a_n)$. The fact that this operation is associative follows from the associativity of addition on $\mathbb{R}$.

If you have studied vector spaces, then you can see that, more generally, the elements of any vector space form a group under vector addition.

4. Consider the binary operation on $\mathbb{R}$ given by $a * b = 2(a + b)$. Then $(\mathbb{R}, *)$ fails to be a group for several reasons. First of all, $*$ is not associative. Secondly, there is no identity element. For suppose e were one. Then we would have $a * e = a$ for *every* a, that is, $2(a + e) = a$. But this implies that $e = -a/2$, and clearly no *one* number can satisfy this equation for *all* a's. Thus there is no identity element e.

However, if we try instead $\mathbb{R} - \{0\}$ with the binary operation $a * b = 2ab$, then we do get a group. For now $*$ is associative, and $1/2$ is an identity element because for any a we have $a * (1/2) = 2a \cdot (1/2) = a$, and similarly $(1/2) * a = 2 \cdot (1/2) \cdot a = a$. What is an inverse for a? If x is to work, then we need $a * x = 1/2 = x * a$, that is, $2ax = 1/2 = 2xa$. Clearly $x = 1/4a$ meets the requirement.

5. Consider the set of 2×2 matrices with real entries, under the binary operation given by matrix multiplication. Is this a group? The only candidate for an identity element is $I = \begin{pmatrix} 1 & 0 \\ 0 & 1 \end{pmatrix}$, because this is the one and only matrix $\begin{pmatrix} e & f \\ g & h \end{pmatrix}$ such that

$$\begin{pmatrix} a & b \\ c & d \end{pmatrix}\begin{pmatrix} e & f \\ g & h \end{pmatrix} = \begin{pmatrix} e & f \\ g & h \end{pmatrix}\begin{pmatrix} a & b \\ c & d \end{pmatrix} = \begin{pmatrix} a & b \\ c & d \end{pmatrix}$$

for all matrices $\begin{pmatrix} a & b \\ c & d \end{pmatrix}$. [Try $\begin{pmatrix} a & b \\ c & d \end{pmatrix} = \begin{pmatrix} 1 & 0 \\ 0 & 1 \end{pmatrix}$!] But then an inverse for $\begin{pmatrix} a & b \\ c & d \end{pmatrix}$ would just be a matrix $\begin{pmatrix} e & f \\ g & h \end{pmatrix}$ such that

$$\begin{pmatrix} a & b \\ c & d \end{pmatrix}\begin{pmatrix} e & f \\ g & h \end{pmatrix} = \begin{pmatrix} e & f \\ g & h \end{pmatrix}\begin{pmatrix} a & b \\ c & d \end{pmatrix} = \begin{pmatrix} 1 & 0 \\ 0 & 1 \end{pmatrix}.$$

Since such a matrix need not exist [try $\begin{pmatrix} a & b \\ c & d \end{pmatrix} = \begin{pmatrix} 0 & 0 \\ 0 & 0 \end{pmatrix}$], we do not have a group.

However, we can attempt to salvage something here by *throwing away* the matrices that do not have inverses. That is, we now let G be the set of all **invertible** 2×2 real matrices. An immediate question is whether matrix multiplication still gives us a binary operation on this smaller set. That is, if A and B are invertible, then is AB invertible? Recalling that matrix multiplication is associative, we see that if A' and B' are inverses for A and B, respectively, then

$$(AB)(B'A') = ((AB)B')A' = (A(BB'))A' = (AI)A' = AA' = I,$$

and similarly

$$(B'A')(AB) = I,$$

so $B'A'$ serves as an inverse for AB.

Thus we do have a binary operation on the restricted set G, and it is associative. There is an identity element in G, because $I = \begin{pmatrix} 1 & 0 \\ 0 & 1 \end{pmatrix}$ is invertible.

Finally, any $A \in G$ has an inverse A' *which is in* G, because the equations $AA' = A'A = I$ tell us that A' is invertible (its inverse is A). Thus G is indeed a group under matrix multiplication. It is called the **general linear group of degree 2 over** $\mathbb{R}$ and denoted by $GL(2, \mathbb{R})$.

If $A = \begin{pmatrix} a & b \\ c & d \end{pmatrix}$ is a matrix, then the number $ad - bc$ is called the **determinant** of A. Determinants are often very useful in working with matrices. For instance, it is easy to check that if $ad - bc \neq 0$, then A is invertible, and the matrix

$$\begin{pmatrix} \dfrac{d}{ad-bc} & \dfrac{-b}{ad-bc} \\[2mm] \dfrac{-c}{ad-bc} & \dfrac{a}{ad-bc} \end{pmatrix}$$

is an inverse for A.

6. Let G consist of all real-valued functions on the real line, with the binary operation given by pointwise addition of functions: If $f, g \in G$ then $f + g$ is the function whose value at any $x \in \mathbb{R}$ is $f(x) + g(x)$.

It is clear that this *does* give us a binary operation on G. To check associativity, we must verify that

$$(f+g)+h = f+(g+h)$$

for all $f, g, h \in G$, and this means that for every $x \in \mathbb{R}$,

$$[(f+g)+h](x) = [f+(g+h)](x), \quad \text{that is,}$$
$$(f+g)(x)+h(x) = f(x)+(g+h)(x), \quad \text{that is,}$$
$$[f(x)+g(x)]+h(x) = f(x)+[g(x)+h(x)],$$

which is true since $f(x)$, $g(x)$, and $h(x)$ are real numbers and addition of real numbers is associative.

The zero function—that is, the function $f_0 \in G$ such that $f_0(x) = 0$ for all $x \in \mathbb{R}$—is an identity element:

$$(f_0 + g) = (g + f_0) = g \quad \text{for all } g \in G.$$

Why? We must check that for any $x \in \mathbb{R}$,

$$(f_0+g)(x) = (g+f_0)(x) = g(x), \quad \text{that is,}$$
$$f_0(x)+g(x) = g(x)+f_0(x) = g(x), \quad \text{that is,}$$
$$0+g(x) = g(x)+0 = g(x),$$

which is true.

If $f \in G$, then $-f$ is an inverse of f, where $-f$ is defined by

$$(-f)(x) = -[f(x)] \quad \text{for all } x \in \mathbb{R}.$$

Why? We have

$$(-f+f)(x)=(-f)(x)+f(x)= -[f(x)]+f(x)=0,$$

and

$$[f+(-f)](x)=f(x)+(-f)(x)=f(x)-[f(x)]=0,$$

for all $x \in \mathbb{R}$. Thus $(-f)+f=f_0$ and $f+(-f)=f_0$, so $(-f)$ is an inverse of f. Thus G is a group with the given operation.

7. Let X be a set, let S be the set of all subsets of X, and consider the binary operation of symmetric difference on S.

This is indeed a binary operation on S, and we have seen that it is associative: $(A \triangle B) \triangle C = A \triangle (B \triangle C)$.

Is there an identity element? In other words, is there $e \in S$ such that $A \triangle e = e \triangle A = A$ for all $A \in S$? Well, $A \triangle e = A$ implies that $A \cap e = \varnothing$, and it also implies that $e \subseteq A$. Thus e would have to be the null set. Let's check that $e = \varnothing$ works:

$$A \triangle \varnothing = (A \cup \varnothing) - (A \cap \varnothing) = A - \varnothing = A,$$

and

$$\varnothing \triangle A = (\varnothing \cup A) - (\varnothing \cap A) = A,$$

so $\varnothing$ is in fact an identity element.

How about inverses? An inverse B of A would have to satisfy

$$A \triangle B = \varnothing = B \triangle A.$$

This implies that $A \subseteq B$ and $B \subseteq A$, so B would have to be A; in fact $A \triangle A = \varnothing$, so A itself *is* an inverse for A.

Thus S forms a group under the operation of symmetric difference. As a matter of notation, the set of subsets of a set X is called the **power set** of X, and denoted by $P(X)$. Thus we have the group $(P(X), \triangle)$.

Finally, we consider some groups obtained from $\mathbb{Z}$.

8. *Additive group of integers modulo n.* Let n be a positive integer and consider the set $\mathbb{Z}_n = \{0, 1, 2, \ldots, n-1\}$. We add the elements of $\mathbb{Z}_n$ modulo n, i.e., we add them as integers and then disregard multiples of n so as to produce an answer that is one of $0, 1, 2, \ldots, n-1$. There is a lemma which we need to make this work. (A *lemma* in mathematics is a preliminary or auxiliary result, not the main result one is aiming at.)

LEMMA 2.1 (The Division Algorithm) If a and n are integers and n is positive, then there exist *unique* integers q and r such that $a = qn + r$ and $0 \leqslant r < n$ (q stands for quotient and r stands for remainder).

PROOF. We prove the existence of q and r, then the uniqueness.

Existence: Let qn be the greatest multiple of n that is $\leqslant a$. If we let $r = a - qn$ then clearly $r \geqslant 0$ and $a = qn + r$, so all we have to check is that $r < n$. But if $r \geqslant n$, then

$$r - n = a - (q+1)n \geqslant 0,$$

so $(q+1)n$ is a multiple of n that is $\leqslant a$, contradicting the choice of qn as the greatest such multiple.

Uniqueness: If $q_1 n + r_1 = q_2 n + r_2$ and $0 \leqslant r_1 < n$, $0 \leqslant r_2 < n$, then

$$q_1 n - q_2 n = r_2 - r_1,$$

so $r_2 - r_1$ is a multiple of n; but $-n < r_2 - r_1 < n$, so $r_2 - r_1 = 0$, so $r_2 = r_1$. Thus $q_1 n = q_2 n$, so $q_1 = q_2$. $\square$

We denote the unique r guaranteed by the lemma by $\bar{a}$, and call it the **remainder of a mod n.**

Observe that two integers a_1 and a_2 have the same remainder mod n iff $a_1 - a_2$ is a multiple of n. When this happens, we say that a_1 and a_2 are **congruent modulo n,** and we write $a_1 \equiv a_2 \pmod{n}$.

Examples Let $n = 7$. Then if $a = 16$, we get $16 = 2 \cdot 7 + 2$, so $q = 2$, $r = 2$, and $\overline{16} = r = 2$. If $a = 35$, we get $35 = 5 \cdot 7 + 0$, so $\overline{35} = 0$.

By the lemma we can unambiguously define a binary operation $\oplus$ on $\mathbb{Z}_n$ by setting $x \oplus y = \overline{x + y}$. For example, if $n = 7$, then $\mathbb{Z}_n = \{0, 1, 2, 3, 4, 5, 6\}$, and $3 \oplus 2 = \overline{3 + 2} = \overline{5} = 5$, $3 \oplus 4 = \overline{3 + 4} = \overline{7} = 0$, and $6 \oplus 3 = \overline{6 + 3} = \overline{9} = 2$.

We claim that this operation turns $\mathbb{Z}_n$ into a group $(\mathbb{Z}_n, \oplus)$.

Associativity: The question is whether $(x \oplus y) \oplus z = x \oplus (y \oplus z)$, that is,

$$\overline{x + y} \oplus z \stackrel{?}{=} x \oplus \overline{y + z},$$

$$\overline{\overline{x + y} + z} \stackrel{?}{=} \overline{x + \overline{y + z}}.$$

Well,

$$\overline{\overline{x + y} + z} = \overline{(x + y) + z}$$

since $\overline{x + y}$ and $(x + y)$ differ by a multiple of n; similarly,

$$\overline{x + \overline{y + z}} = \overline{x + (y + z)}.$$

So all we have to check is whether $\overline{(x + y) + z}$ equals $\overline{x + (y + z)}$. But this is obviously so, since $(x + y) + z = x + (y + z)$.

Identity: The identity element is 0. Why? $0 \oplus x = \overline{0+x} = \bar{x} = x$ if $0 \leqslant x \leqslant n-1$, and, similarly, $x \oplus 0 = x$.

Inverses: We know that 0 is an inverse for 0, since $0 \oplus 0 = \overline{0+0} = \bar{0} = 0$, the identity element. Now if $x \neq 0$, then $x \in \{1, 2, \ldots, n-1\}$, so $n-x \in \{1, 2, \ldots, n-1\}$, and we see that $n-x$ is an inverse of x:

$$x \oplus (n-x) = \overline{x+n-x} = \bar{n} = 0,$$

and, similarly, $(n-x) \oplus x = 0$.

For example, in $(\mathbb{Z}_7, \oplus)$, 1 is an inverse for 6, and 3 is an inverse for 4. In $(\mathbb{Z}_{13}, \oplus)$, 5 is an inverse for 8.

The group $(\mathbb{Z}_n, \oplus)$ is called the **additive group of integers mod n**. Notice that, for $n=1$, we have $\mathbb{Z}_1 = \{0\}$, so $(\mathbb{Z}_1, \oplus)$ is a group with only one element in it. In general, any group having only one element is called **trivial**. If x is any object whatsoever (e.g., $x = $ Whistler's Mother), then we get a trivial group $(\{x\}, *)$ by defining $x * x = x$.

EXERCISES

2.1 Which of the following are groups? Why?

a) $\mathbb{R}^+$ under addition

b) The set $3\mathbb{Z}$ of integers that are multiples of 3, under addition

c) $\mathbb{R} - \{0\}$ under the operation $a * b = |ab|$

d) The set $\{1, -1\}$ under multiplication

e) The subset of $\mathbb{Q}$ consisting of all positive rationals that have rational square roots, under multiplication

f) The set of all pairs (x, y) of real numbers, under the operation $(x, y) * (z, w) = (x+z, y-w)$

g) The set of all pairs (x, y) of real numbers such that $y \neq 0$, under the operation $(x, y) * (z, w) = (x+z, yw)$

h) $\mathbb{R} - \{1\}$, under the operation $a * b = a + b - ab$

i) $\mathbb{Z}$, under the operation $a * b = a + b - 1$

2.2 a) Of those examples in Exercise 2.1 that are groups, which are abelian?

b) Which of the groups in Examples 1–8 on pp. 17–20 are abelian?

2.3 Let X be a set and let $P(X)$ be the power set of X. Does $P(X)$ with the binary operation $A * B = A \cap B$ form a group? How about $P(X)$ with the binary operation $A * B = A \cup B$?

2.4 The operation in a finite group can be specified by writing down a table (see Exercise 1.9). Write down the tables for the following.

a) $(\mathbb{Z}_4, \oplus)$

b) $(\mathbb{Z}_5, \oplus)$

c) $(\mathbb{Z}_6, \oplus)$

2.5 The following table defines a binary operation on the set $S = \{a,b,c\}$.

*	a	b	c
a	a	b	c
b	b	b	c
c	c	c	c

Is $(S, *)$ a group?

2.6 The following table defines a binary operation on the set $S = \{a,b,c\}$.

*	a	b	c
a	a	b	c
b	b	a	c
c	c	b	a

Is $(S, *)$ a group?

2.7 Let $S = \{a,b\}$. Write down a table that defines a binary operation $*$ on S such that $(S, *)$ is a group. Show that your table works.

2.8 Let G be the set of all real-valued functions f on the real line which have the property that $f(x) \neq 0$ for all $x \in \mathbb{R}$. Define the product $f \times g$ of two functions f, g in G by

$$(f \times g)(x) = f(x)g(x) \quad \text{for all } x \in \mathbb{R}.$$

With this operation, does G form a group? Prove or disprove.

2.9 a) Show that for 2×2 matrices A and B,

determinant of $AB = $ (determinant of A)(determinant of B).

b) Show that a 2×2 matrix A is in $GL(2, \mathbb{R})$ iff the determinant of A is not 0.

c) Use the results of (a) and (b) to give another proof that $GL(2, \mathbb{R})$ is a group under matrix multiplication.

2.10 Let G be the set of all 2×2 matrices $\begin{pmatrix} a & b \\ -b & a \end{pmatrix}$, where $a,b \in \mathbb{R}$ and $a^2 + b^2 \neq 0$. Show that G forms a group under matrix multiplication.

2.11 Let G be the set of all 2×2 matrices $\begin{pmatrix} a & 0 \\ 0 & b \end{pmatrix}$, where a and b are nonzero real numbers. Show that G forms a group under matrix multiplication.

2.12 Let a,b,c,d be elements of a group $(G, *)$. Show that

$$(a*b)*(c*d) = ((a*b)*c)*d = (a*(b*c))*d$$
$$= a*((b*c)*d) = a*(b*(c*d)).$$

2.13 Let $a_1, a_2, \ldots, a_n$ be elements of a group G. Show that

$$a_1 * a_2 * a_3 * \cdots * a_n$$

has an unambiguous meaning in the sense that no matter how we insert parentheses into the expression to indicate the order in which the multiplications are carried out, we always get the same result. [*Suggestion:* Show that any insertion of parentheses gives the same answer as

$$a_1 * (a_2 * (a_3 * (a_4 * \cdots * (a_{n-1} * a_n) \ldots))).$$

To do this, use induction on n.]

FUNDAMENTAL THEOREMS ABOUT GROUPS

If we hope to get anywhere in working with groups, there are certain fundamental facts about their behavior that we must master at the outset. The operation * in a group comes to us endowed only with the properties given to it by the group axioms. Everything else must follow from these, and our first task is to use the axioms to set down some basic rules of operation that enable us to carry out with ease at least some elementary calculations.

First, a convention: We usually call the operation in a group "multiplication"; but very often the operation is called "addition" if the group happens to be abelian.

THEOREM 3.1 (Uniqueness of the identity element) If $(G, *)$ is a group, then there is only one identity element in G.

PROOF. We must establish that if e and e_1 are two elements of G both of which satisfy the defining property of an identity element in G, then in fact $e = e_1$. That is, we *assume* that $x * e = e * x = x$ for *all* x in G *and* $x * e_1 = e_1 * x = x$ for *all* x in G, and we then proceed to show that e must equal e_1.

By the assumption on e, we have in particular (taking x to be e_1)
$$e_1 * e = e * e_1 = e_1.$$
By the assumption on e_1, we have (taking x to be e)
$$e * e_1 = e_1 * e = e.$$
Thus we have $e_1 = e * e_1 = e$, and the proof is complete. $\square$

THEOREM 3.2 (Uniqueness of inverses) If $(G, *)$ is a group and x is any element of G, then x has only one inverse in G.

PROOF. We must show that if both y_1 and y_2 satisfy the definition of an inverse of x, that is, if $x * y_1 = y_1 * x = e$ and $x * y_2 = y_2 * x = e$, then in fact $y_1 = y_2$.

25

We will take some of the given information and use it to derive an equation which has y_1 on one side and y_2 on the other. It may leap to your eye, for example, that

$$x * y_1 = x * y_2,$$

because both sides are e. Once this is seen, all that remains is to get rid of the x's. We can do this by using more of the given information, namely the fact that $y_1 * x = e$. We multiply both sides by y_1:

$$y_1 * (x * y_1) = y_1 * (x * y_2).$$

By using associativity, we get from this

$$(y_1 * x) * y_1 = (y_1 * x) * y_2,$$
$$e * y_1 = e * y_2,$$
$$y_1 = y_2,$$

as desired. $\square$

The proof is finished, but we are going to do it again in a slightly different way to illustrate the fact that there are often several different ways to prove something. Suppose, for example, that the equation $x * y_1 = x * y_2$ did *not* leap to your eye, and that you just took one of the equations given, say

$$y_1 * x = e,$$

to start with. If you want to get from this an equation with y_1 on one side and y_2 on the other, then certainly you observe that y_1 is already on the left, and we can get y_2 on the right by multiplying both sides by y_2:

$$(y_1 * x) * y_2 = e * y_2$$
$$= y_2.$$

Now if only we could get rid of the x and y_2 on the left-hand side, we'd be done; but we know from the equations we had to start with that $x * y_2 = e$, so we rewrite the left side by using associativity, so as to bring $x * y_2$ into play:

$$y_1 * (x * y_2) = y_2,$$
$$y_1 * e = y_2,$$
$$y_1 = y_2,$$

and we are done. $\square$

We emphasize the significance of what we have just proved twice: The group axioms simply assert that for any x, there must exist an inverse; our theorem says that once you know you've got a group, any x has precisely *one* inverse.

Example Let $(G, *)$ be $GL(2, \mathbb{R})$. Then by applying our general result in this specific case, we conclude that if $\begin{pmatrix} a & b \\ c & d \end{pmatrix}$ is an invertible matrix, then there is only *one* matrix $\begin{pmatrix} e & f \\ g & h \end{pmatrix}$ such that

$$\begin{pmatrix} a & b \\ c & d \end{pmatrix}\begin{pmatrix} e & f \\ g & h \end{pmatrix} = \begin{pmatrix} e & f \\ g & h \end{pmatrix}\begin{pmatrix} a & b \\ c & d \end{pmatrix} = \begin{pmatrix} 1 & 0 \\ 0 & 1 \end{pmatrix}.$$

This fact can of course be established directly, by working with systems of linear equations in two variables. But one is struck by the economy and elegance—the "cleanness"—of the proof we have obtained by viewing the collection of all invertible 2×2 matrices as a group.

Henceforth we will usually denote the unique inverse of x by x^{-1}. When we are dealing with an abelian group and referring to the group operation as addition, however, we will sometimes denote the inverse of x by $-x$. For example, in $GL(2, \mathbb{R})$ we write

$$\begin{pmatrix} 3 & 5 \\ 4 & 7 \end{pmatrix}^{-1} = \begin{pmatrix} 7 & -5 \\ -4 & 3 \end{pmatrix},$$

and in $(\mathbb{Z}_7, \oplus)$ we write $-3 = 4$.

The next result will enable us to conclude that no two distinct elements of a group G can have the same inverse.

THEOREM 3.3 If $(G, *)$ is a group, then for any $x \in G$ we have $(x^{-1})^{-1} = x$.

PROOF. Since x^{-1} is the inverse of x, we know that $x^{-1} * x = x * x^{-1} = e$. By these equations, x satisfies the definition of $(x^{-1})^{-1}$, so $x = (x^{-1})^{-1}$ by the uniqueness of inverses. $\square$

That was slick, but let's do it again in a slightly different way. We know that $x^{-1} * x = e$. We want to get from this to an equation with x on one side and $(x^{-1})^{-1}$ on the other. We need to get rid of the x^{-1} on the left side, so let's multiply both sides by $(x^{-1})^{-1}$:

$$(x^{-1})^{-1} * (x^{-1} * x) = (x^{-1})^{-1} * e$$
$$((x^{-1})^{-1} * x^{-1}) * x = (x^{-1})^{-1} \quad \text{(using associativity on the left)}$$
$$e * x = (x^{-1})^{-1}$$
$$x = (x^{-1})^{-1}. \quad \square$$

Example Let $G=(\mathbb{Z}, +)$. Then for this example the theorem says that $-(-n)=n$ for any integer n.

Now suppose x,y are elements of a group $(G,*)$ and $x^{-1}=y^{-1}$. Then by taking inverses on both sides we get $(x^{-1})^{-1}=(y^{-1})^{-1}$, so, by the theorem, $x=y$. Thus, as promised, if two elements have the same inverse then they must in fact be the same element. It is possible to prove this without reference to the preceding theorem; see Exercise 3.8(b).

Next we examine the inverse of a product.

THEOREM 3.4 If $(G,*)$ is a group and $x, y \in G$, then

$$(x*y)^{-1}=y^{-1}*x^{-1}.$$

PROOF.

$$(x*y)*(y^{-1}*x^{-1})=x*(y*(y^{-1}*x^{-1}))$$

$$=x*((y*y^{-1})*x^{-1})=x*(e*x^{-1})=x*x^{-1}=e,$$

and similarly we can show that $(y^{-1}*x^{-1})*(x*y)=e$. (Do it!) Thus the element $y^{-1}*x^{-1}$ satisfies the conditions that define $(x*y)^{-1}$, and since we already know that inverses are unique, this implies that $y^{-1}*x^{-1}$ and $(x*y)^{-1}$ must be the same element. $\square$

It is worth emphasizing the reversal of order in the above result. In general, it is *not* true that $(x*y)^{-1}$ is $x^{-1}*y^{-1}$. This does, of course, hold true in *abelian* groups, for then $x^{-1}*y^{-1}$ is the same thing as $y^{-1}*x^{-1}$; in fact, the equation $(x*y)^{-1}=x^{-1}*y^{-1}$ holds for all x, y in a group G if and only if G is abelian (Exercise 3.9).

It is somewhat bothersome to have to check both the conditions $(x*y)*(y^{-1}*x^{-1})=e$ and $(y^{-1}*x^{-1})*(x*y)=e$ in the above theorem and, in fact, it is not hard to show that it is really sufficient to check either one of them.

THEOREM 3.5 Let $(G,*)$ be a group and let $x, y \in G$. Suppose that *either* $x*y=e$ *or* $y*x=e$. Then y is x^{-1}.

PROOF. Suppose that $x*y=e$. We wish to solve this equation for y, so let's multiply both sides by x^{-1}:

$$x^{-1}*(x*y)=x^{-1}*e.$$

Thus

$$\left(x^{-1} * x\right) * y = x^{-1},$$
$$e * y = x^{-1},$$
$$y = x^{-1}.$$

A similar argument shows that $y * x = e$ is also by itself sufficient to guarantee that $y = x^{-1}$. (Do it!) □

The same solving of an equation proves the more general

THEOREM 3.6 (Cancellation laws) Let $(G, *)$ be a group and let x, y, $z \in G$. Then:

i) if $x * y = x * z$, then $y = z$; and

ii) if $y * x = z * x$, then $y = z$.

The proof is left as an exercise. Part (i) is called the **left cancellation law**, and Part (ii) the **right cancellation law**.

We close this section by giving another formulation of the axioms for a group which is equivalent to our original definition, but somewhat simpler to work with in establishing that some system is in fact a group.

THEOREM 3.7 Let G be a set and $*$ an associative binary operation on G. Assume that there is an element $e \in G$ such that $x * e = x$ for all $x \in G$, and assume that for any $x \in G$ there exists an element y in G such that $x * y = e$. Then $(G, *)$ is a group.

The element e is called a **right identity**, and the element y associated to x is called a **right inverse** of x. In order to prove the theorem we have to show that G satisfies all the axioms for a group, and since we have assumed that $*$ is a binary operation on G and $*$ is associative, we have only to verify that e is, in fact, also a **left identity**, that is, $e * x = x$ for all $x \in G$; and that a right inverse y of x is also, in fact, a **left inverse** of x, that is, $y * x = e$.

PROOF OF THE THEOREM. First we show that $e * x = x$ for all $x \in G$. Let us do the proof backwards by trying to obtain some equations that we know would yield $e * x = x$. Let x' denote a right inverse of x. Certainly it would be enough to have

$$(e * x) * x' = x * x', \qquad [3.1]$$

for then we could multiply both sides by a right inverse of x'. But having [3.1] is the same thing as having

$$e * (x * x') = x * x',$$

by associativity, and this is the same thing as having

$$e * e = e,$$

by the definition of the right inverse x'. Certainly we *know* that $e * e = e$, because $x * e = x$ for any $x \in G$.

We have proved $e * x = x$, because each successive equation in our proof implied the one before it, so that when we finally arrived at a true equation, its truth implied the truth of all the previous equations. It is crucial to realize that in this situation it would not have sufficed to have each equation implying the one *after* it. In other words, if we want to prove $e * x = x$, it suffices to show that $e * x = x$ is implied by the true statement $e * e = e$, but it would not be enough to show that $e * x = x$ *implies* $e * e = e$. (A simple example: The false statement "$-1 = 1$" implies the true statement "$1 = 1$," as we see by squaring both sides; but that doesn't prove $-1 = 1$.)

Now let's finish the proof of Theorem 3.7 by showing that a right inverse x' of x is also a left inverse of x, that is, $x' * x = e$. We know that there is some $(x')'$ such that $x' * (x')' = e$, and it will suffice to show that $x = (x')'$. Now from

$$x * x' = e$$

we get

$$(x * x') * (x')' = e * (x')',$$

and since we know that e is a left identity, this yields $x = (x')'$. □

An analogous proof shows that assuming associativity and the existence of a left identity and left inverses is also sufficient to guarantee a group. It should be observed, however, that associativity plus the existence of a right identity and left inverses (or a left identity and right inverses) is *not* enough.

Example Consider the set $\mathbb{Z}$ with the binary operation given by

$$x * y = x.$$

It is easy to check that $*$ is an associative binary operation on $\mathbb{Z}$ and that 1 is a right identity element and also a left inverse for every element of $\mathbb{Z}$. However $(\mathbb{Z}, *)$ is not a group since, for example, there is no two-sided identity element in $(\mathbb{Z}, *)$.

Since the axioms in Theorem 3.7 are manifestly simpler than those in our definition of a group, you may wonder why we didn't use the simpler version as the definition and then show that the stronger axioms follow. We could have done so, but we decided against it in favor of emphasizing the fact that in a group the identity element and inverses work from both sides.

EXERCISES

3.1 In $(\mathbb{Z}_{12}, \oplus)$, solve the equation $2 \oplus x \oplus 7 = 1$ for x.

3.2 Let $X = \{1, 2, 3, 4, 5, 6, 7, 8, 9, 10\}$. In $(P(X), \triangle)$, consider the elements $A = \{1, 4, 5, 7, 8\}$ and $B = \{2, 4, 6\}$, and solve $A * x = B$ for x.

3.3 Find elements A, B, C of $GL(2, \mathbb{R})$ such that $AB = BC$ but $A \neq C$.

3.4 Let g be an element of a group $(G, *)$ such that for some one element $x \in G$, $x * g = x$. Show that $g = e$.

3.5 If $(G, *)$ is a group and $x, y, z \in G$, then we can unambiguously write $x * y * z$ to denote either $(x * y) * z$ or $x * (y * z)$, since, by associativity, these are the same element. Show that

$$(x * y * z)^{-1} = z^{-1} * y^{-1} * x^{-1}.$$

3.6 Prove the cancellation laws (Theorem 3.6).

3.7 Let G be a finite group, and consider the multiplication table for G, i.e., the table that gives the binary operation of G (see Exercise 1.9). Show that every element of G occurs precisely once in each row of the table and precisely once in each column.

3.8 Use the cancellation laws to give alternative proofs of:

a) Theorem 3.1;

b) the fact that if $x^{-1} = y^{-1}$ then $x = y$.

3.9 Let $(G, *)$ be a group. Show that $(G, *)$ is abelian iff

$$(x * y)^{-1} = x^{-1} * y^{-1} \quad \text{for all } x, y \in G.$$

3.10 Let $(G, *)$ be a group and let g be some fixed element of G. Show that $G = \{ g * x \mid x \in G \}$.

3.11 Let $(G, *)$ be a group such that $x^2 = e$ for all $x \in G$. Show that $(G, *)$ is abelian. (Here x^2 means $x * x$.)

3.12 Let $(G, *)$ be a group. Show that $(G, *)$ is abelian iff $(x * y)^2 = x^2 * y^2$ for all x, y in G.

3.13 Let G be a set and let $*$ be an associative binary operation on G. Assume that there exists a left identity element in G and that every element in G has a left inverse. Prove that $(G, *)$ is a group.

3.14 Let G be a nonempty set and let $*$ be an associative binary operation on G. Assume that for any elements a, b in G, we can find $x \in G$ such that $a * x = b$, and we can find y such that $y * a = b$. Show that $(G, *)$ is a group.

3.15 Let G be a nonempty set and let $*$ be an associative binary operation on G. Assume that both the left and right cancellation laws hold in $(G, *)$. Assume moreover that G is *finite*. Show that $(G, *)$ is a group. (*Suggestions*: Show first that there exists a right identity element. To do this, suppose that $a_1, a_2, \ldots, a_n$ are all the elements of G, and consider the set

$$\{a_1 * a_1, a_1 * a_2, \ldots, a_1 * a_n\}.$$

Show that the elements of this set are all distinct and that, therefore, one of them must be a_1. If, say, $a_1 * a_j = a_1$, then a_j is your candidate for a right identity element. You must still show that $x * a_j = x$ for *every* $x \in G$. With this done, all that remains is to show that every element of G has a right inverse.)

3.16 Give an example to show that if the assumption that G is finite is omitted from Exercise 3.15, then the conclusion need no longer follow, i.e., $(G, *)$ need not be a group.

POWERS OF AN ELEMENT;
CYCLIC GROUPS

Before going any further in our investigation of groups, we pause to stream-line our notation. You have probably already started to get tired of writing $*$ every time you want to indicate the operation in a group $(G, *)$. It is common practice to avoid this encumbrance by writing xy in place of $x * y$, so long as no confusion can arise. For example, the equation

$$(x * y)^{-1} = y^{-1} * x^{-1}$$

is usually written

$$(xy)^{-1} = y^{-1} x^{-1},$$

and the associative law

$$(x * y) * z = x * (y * z)$$

is written

$$(xy)z = x(yz).$$

In keeping with this simplification, we will usually refer to an abstract group as G, rather than $(G, *)$.

   In discussing concrete examples, we continue to use whatever notation is appropriate. For example, if we are talking about $(\mathbb{Z}, +)$ we write $x + y$. You are also reminded that the additive notation is very commonly used in discussing any *abelian* group.

   Another economy in notation is achieved by taking advantage of the associative law in order to eliminate parentheses. For example, we can unambiguously write xyz to denote either $(xy)z$ or $x(yz)$, since these two elements are the same. Similarly, if $x_1, \ldots, x_n$ are elements of a group, then $x_1 x_2 x_3 \cdots x_n$ has an unambiguous meaning; no matter how we insert parentheses into the expression, the resulting product always equals

$$x_1(x_2(x_3(\cdots(x_{n-1}x_n)\cdots))).$$

Verifying this for yourself is a good way to check your understanding of the associative law. (See Exercise 2.13.)

A word of caution: There are times when parentheses cannot be omitted without changing the meaning of an expression. For example, $(xy)^{-1}$ is not in general the same thing as xy^{-1}.

Now let x be an element of a group G. We define the powers x^n of x (for $n \in \mathbb{Z}$) as follows:

$$x^0 = e;$$
$$x^n = xxx \cdots x \ (n \ \textit{factors}), \quad \text{if } n > 0;$$
$$x^{-n} = (x^{-1})^n = x^{-1}x^{-1}x^{-1} \cdots x^{-1}, \quad \text{if } n > 0.$$

Here are the rules for working with exponents.

THEOREM 4.1 Let G be a group and let $x \in G$. Let m, n be integers. Then:

i) $x^m x^n = x^{m+n}$;

ii) $(x^n)^{-1} = x^{-n}$;

iii) $(x^m)^n = x^{nm} = (x^n)^m$.

PROOF. i) First suppose that m and n are both positive. Then

$$x^m x^n = \underbrace{xx \cdots x}_{m \ \textit{factors}} \cdot \underbrace{xx \cdots x}_{n \ \textit{factors}} = \underbrace{xx \cdots x}_{m+n \ \textit{factors}} \quad \text{(by associativity)},$$

and this is x^{m+n}. Next, if m and n are both negative, say $m = -r$ and $n = -s$, then

$$x^m x^n = x^{-r}x^{-s} = (x^{-1})^r(x^{-1})^s = (x^{-1})^{r+s} \quad \text{(by the first case)},$$

and this is $x^{-(r+s)}$, that is, x^{m+n}. If $m = -r < 0$ and $n > 0$, then if $r > n$ we have

$$x^m x^n = (x^{-1})^r x^n = (x^{-1})^{r-n} = x^{-(r-n)} = x^{-r+n} = x^{m+n}.$$

The remaining cases can be treated similarly, and are left to the reader.

(ii) and (iii) are exercises. $\square$

We observe that if we are writing the group operation additively, then x^2 means $x + x$, x^3 means $x + x + x$, and so on. In this context, we usually write nx in place of x^n; then (i) above becomes $mx + nx = (m + n)x$, (ii) becomes $-(nx) = (-n)x$, and (iii) becomes $n(mx) = (nm)x = m(nx)$.

DEFINITIONS If G is a group and $x \in G$, then x is said to be **of finite order** if there exists a positive integer n such that $x^n = e$. If such an integer exists, then the smallest positive n such that $x^n = e$ is called the **order** of x and denoted by $o(x)$. If x is not of finite order, then we say that x is **of infinite order** and write $o(x) = \infty$.

Examples

1. Let G be $(\mathbb{Z}_3, \oplus)$. Then $o(1) = 3$, since
    $$1 \neq 0, \ 1 \oplus 1 \neq 0, \quad \text{and} \quad 1 \oplus 1 \oplus 1 = 0.$$
2. Let G be $(\mathbb{Z}, +)$. Then $o(1) = \infty$, since
    $$1 \neq 0, \quad 1 + 1 \neq 0, \quad 1 + 1 + 1 \neq 0, \quad \text{etc.}$$
3. Let G be $(\mathbb{Q}^+, \cdot)$. Then $o(2) = \infty$, since
    $$2 \neq 1, \quad 2^2 \neq 1, \quad 2^3 \neq 1, \quad 2^4 \neq 1, \quad \text{etc.}$$
4. Let G be $GL(2, \mathbb{R})$. Then $o\left(\begin{pmatrix} -1 & 0 \\ 0 & -1 \end{pmatrix} \right) = 2$ since

$$\begin{pmatrix} -1 & 0 \\ 0 & -1 \end{pmatrix} \neq \begin{pmatrix} 1 & 0 \\ 0 & 1 \end{pmatrix}, \quad \text{but} \quad \begin{pmatrix} -1 & 0 \\ 0 & -1 \end{pmatrix}^2 = \begin{pmatrix} 1 & 0 \\ 0 & 1 \end{pmatrix}.$$

Since the notion of "order of an element" is defined in terms of integers, it is not surprising that one needs some information on integers in order to investigate its properties. We include this material at this point as a change of pace.

If m, n are integers, not both zero, then (m, n) denotes the **greatest common divisor** (g.c.d.) of m and n. This is by definition the largest integer d that divides both m and n. [If $m = n = 0$, then (m, n) doesn't exist because *every* integer divides 0: $0 = k \cdot 0$ for any k.] It is clear that $(m, n) = (|m|, |n|)$, so that in what follows we can assume that m and n are nonnegative integers, at least one of which is not zero.

There is a process called the **Euclidean algorithm** which enables us to find (m, n) by doing some arithmetic. Say $n \leqslant m$; then we can find unique integers q and r such that
$$m = qn + r \quad \text{and} \quad 0 \leqslant r < n.$$
Now any integer divides m and n if and only if it divides n and r, so the greatest integer that divides m and n is the same as the greatest integer that divides n and r, that is,
$$(m, n) = (n, r).$$
The good thing about knowing this is that we have traded in the pair m, n for the pair n, r and in so doing we have replaced the greater of m, n by r, which is smaller than the smaller of m, n. If $r = 0$, then clearly $(m, n) = n$ and we are done. If $r \neq 0$, then we repeat our trick and find q_1 and r_1 such that
$$n = q_1 r + r_1, \quad \text{where } 0 \leqslant r_1 < r.$$
Then
$$(m, n) = (n, r) = (r, r_1),$$

and we have traded in n for r_1, which is even smaller than r. If $r_1 = 0$, then $(m, n) = r$. If $r_1 \neq 0$, we find q_2 and r_2 such that

$$r = q_2 r_1 + r_2$$

and $0 \leqslant r_2 < r_1$. Then

$$(m, n) = (n, r) = (r, r_1) = (r_1, r_2),$$

and if $r_2 = 0$, $(m, n) = r_1$. If $r_2 \neq 0$, then we continue the process, obtaining a sequence of remainders $r > r_1 > r_2 > r_3 > \ldots$, where each $r_i \geqslant 0$. By the well-ordering principle, such a decreasing sequence of nonnegative integers cannot go on forever, so some r_i must eventually be 0. If so, then

$$(m, n) = (n, r_1) = (r_1, r_2) = \cdots = (r_{i-1}, r_i) = (r_{i-1}, 0) = r_{i-1}.$$

Thus (m, n) is the last nonzero remainder arising from our repeated divisions.

Example Let's find $(1251, 1976)$:

$$1976 = 1251 \cdot 1 + 725,$$
$$1251 = 725 \cdot 1 + 526,$$
$$725 = 526 \cdot 1 + 199,$$
$$526 = 199 \cdot 2 + 128,$$
$$199 = 128 \cdot 1 + 71,$$
$$128 = 71 \cdot 1 + 57,$$
$$71 = 57 \cdot 1 + 14,$$
$$57 = 14 \cdot 4 + 1,$$
$$14 = 1 \cdot 14 + 0.$$

Here $r = 725$, $r_1 = 526$, $r_2 = 199$, $r_3 = 128$, $r_4 = 71$, $r_5 = 57$, $r_6 = 14$, $r_7 = 1$, $r_8 = 0$. Thus $(1251, 1976) = r_7 = 1$. We indicate the fact that the g.c.d. is 1 by saying that 1251 and 1976 are **relatively prime**.

Actually our interest in this process is not so much that it enables us to find (m, n), but that it allows us to establish the following fact.

THEOREM 4.2 If m and n are integers, not both zero, then there exist integers x and y such that

$$mx + ny = (m, n).$$

Thus the g.c.d. of m and n can be written as a "linear combination" of m and n, with integer coefficients.

The utility of this information will become clear in a moment.

PROOF OF THE THEOREM. We write down the steps in the calculation of (m,n) by the Euclidean algorithm, and then use them in reverse order. We have

$$m = qn + r,$$
$$n = q_1 r + r_1,$$
$$r = q_2 r_1 + r_2,$$
$$r_1 = q_3 r_2 + r_3,$$
$$\vdots$$
$$r_{i-4} = q_{i-2} r_{i-3} + r_{i-2},$$
$$r_{i-3} = q_{i-1} r_{i-2} + r_{i-1}, \quad (r_{i-1} \neq 0)$$
$$r_{i-2} = q_i r_{i-1} + 0,$$

so $r_{i-1} = (m,n)$. Now the next-to-last step can be written as

$$r_{i-1} = 1 \cdot r_{i-3} - q_{i-1} \cdot r_{i-2}, \qquad [4.1]$$

so (m,n) is written as a linear combination of r_{i-3} and r_{i-2}. The preceding step $(r_{i-4} = q_{i-2} r_{i-3} + r_{i-2})$ can be used to replace r_{i-2} by $r_{i-4} - q_{i-2} r_{i-3}$ in Eq. [4.1], resulting in an expression for (m,n) as a linear combination of r_{i-4} and r_{i-3}. Using all the equations from the Euclidean algorithm in reverse order, we eventually arrive at an expression for (m,n) as a linear combination of m and n. $\square$

Example We find x and y such that $(1251, 1976) = 1251x + 1976y$. Referring back to our calculation of $(1251, 1976) = 1$, we get:

$$
\begin{aligned}
1 &= 1 \cdot 57 - 4 \cdot 14 \\
&= 1 \cdot 57 - 4(71 - 57) = -4 \cdot 71 + 5 \cdot 57 \\
&= -4 \cdot 71 + 5(128 - 71) = -9 \cdot 71 + 5 \cdot 128 \\
&= 5 \cdot 128 - 9(199 - 128) = 14 \cdot 128 - 9 \cdot 199 \\
&= -9 \cdot 199 + 14(526 - 199 \cdot 2) = -37 \cdot 199 + 14 \cdot 526 \\
&= 14 \cdot 526 - 37(725 - 526) = 51 \cdot 526 - 37 \cdot 725 \\
&= -37(725) + 51(1251 - 725) = -88 \cdot 725 + 51 \cdot 1251 \\
&= 51 \cdot 1251 - 88(1976 - 1251) = 139 \cdot 1251 - 88 \cdot 1976.
\end{aligned}
$$

Thus we can take $x = 139$ and $y = -88$.

After *that*, one should be in a good frame of mind to appreciate some good clean abstraction, but before returning to groups we derive a consequence of the last result that will be useful.

THEOREM 4.3 (Euclid) If r, s, t are integers, r divides st, and $(r,s) = 1$, then r divides t.

PROOF. Since $(r,s) = 1$, there exist integers x and y such that

$$rx + sy = 1.$$

Multiplying both sides of this equation by t yields

$$rxt + syt = t, \quad \text{or}$$
$$r(xt) + st(y) = t.$$

Manifestly, r divides $r(xt)$; and r divides $st(y)$ since it divides st by assumption. Thus r divides the sum of $r(xt)$ and $st(y)$, i.e., r divides t, as claimed. $\square$

There are many simple results in number theory which never lose their charm, and that's one of them.

Back to groups.

THEOREM 4.4 Let G be a group and $x \in G$.

i) $o(x) = o(x^{-1})$.

ii) If $o(x) = n$ and $x^m = e$, then n divides m.

iii) If $o(x) = n$ and $(m, n) = d$, then $o(x^m) = n/d$.

PROOF. The proof of part i) is left as an exercise.

ii) We have $x^m = e$ and we seek to make something of the fact that n is the *smallest* positive integer such that $x^n = e$. Write $m = qn + r$, where $0 < r < n$. Then $x^m = e$ becomes

$$x^{qn+r} = e$$
$$x^{qn}x^r = e$$
$$(x^n)^q x^r = e.$$

But $x^n = e$, so the last equation becomes

$$x^r = e.$$

But r is *smaller than* n, so this is impossible unless $r = 0$. Thus $m = qn + 0$, so n divides m.

iii) Here we use our information about greatest common divisors. We must show that n/d is the smallest positive integer k such that $(x^m)^k = e$. First of all,

$$(x^m)^{n/d} = x^{m \cdot (n/d)} = x^{(m/d) \cdot n} = (x^n)^{m/d} = e^{m/d} = e,$$

since $o(x) = n$. Now suppose $k > 0$ is *smaller* than n/d and $(x^m)^k = e$. We will show that n/d divides k, a contradiction. We have $x^{mk} = e$, so by part (ii) we know that n divides mk, which implies that n/d divides $(m/d) \cdot k$. Since $(m/d, n/d) = 1$ (why?), this implies that n/d divides k (by Theorem 4.3), which is patently absurd since k is a positive integer smaller than n/d. $\square$

We will use these results on the order of an element in Section 5, to help us obtain some results about what are known as *cyclic* groups. For now, we will just introduce cyclic groups.

We remarked that the study of abstract group theory evolved from the study of specific examples. The abstract concept was formulated in an effort to bring together certain concrete cases. Once this was done, however, there was, of course, a new problem. How far-reaching was the abstract concept? What kinds of groups were there *other* than those that motivated the abstraction?

A central goal of group theory is to classify all groups, i.e., to see what kinds of groups there are. One would like to start with the easiest groups. It turns out that these are the cyclic groups—those groups that are just the set of powers of some one element.

DEFINITIONS A group G is called **cyclic** if there is an element $x \in G$ such that $G = \{x^n | n \in \mathbb{Z}\}$; x is then called a **generator** for G.

It will be convenient to have a more compact notation for the set $\{x^n | n \in \mathbb{Z}\}$. We will denote it by $\langle x \rangle$. Thus G is cyclic with x as a generator iff $G = \langle x \rangle$.

In additive notation, $\langle x \rangle = \{nx | n \in \mathbb{Z}\}$.

Examples

1. $(\mathbb{Z}_1, \oplus)$ is the trivial group $\{0\}$ consisting of just an identity element. Clearly, then, $(\mathbb{Z}_1, \oplus) = \langle 0 \rangle$.

2. If $n > 2$, then $(\mathbb{Z}_n, \oplus) = \langle 1 \rangle$, for the powers

$$1, 1 \oplus 1, 1 \oplus 1 \oplus 1, \ldots, \underbrace{1 \oplus 1 \oplus 1 \oplus \cdots \oplus 1}_{n \text{ terms}}$$

   exhaust $(\mathbb{Z}_n, \oplus)$.

3. $(\mathbb{Z}, +)$ is cyclic with generator 1, that is, $(\mathbb{Z}, +) = \langle 1 \rangle$. In this case we have to use all the powers of the generator to get all of the group: $0, 1, -1, 1 + 1, -1 - 1, 1 + 1 + 1, -1 - 1 - 1$, and so on.

4. $(\mathbb{Q}, +)$ is *not* cyclic. For clearly 0 is not a generator, and if $q \neq 0$ then we can easily exhibit rational numbers that are not in $\langle q \rangle = \{nq | n \in \mathbb{Z}\}$. An example is $q/2$.

It should be made explicit that the powers of an element need not all be distinct. In fact, we have the following result:

THEOREM 4.5 Let $G = \langle x \rangle$. If $o(x) = \infty$, then $x^j \neq x^k$ for $j \neq k$, and consequently G is infinite. If $o(x) = n$, then $x^j = x^k$ iff $j \equiv k \pmod{n}$, and consequently the distinct elements of G are $e, x, x^2, \ldots, x^{n-1}$.

PROOF. Suppose that $j \neq k$ and $x^j = x^k$. If, say, $j > k$, then we obtain $x^{j-k} = e$, and $j - k > 0$, so x has finite order. This proves the first statement.

For the second, suppose that $o(x) = n$. Then $x^j = x^k$ iff $x^{j-k} = e$ iff (by Theorem 4.4 ii) n divides $j - k$ iff $j \equiv k \pmod{n}$. $\square$

DEFINITION The **order** of a group G, denoted by $|G|$, is the number of elements in G.

Theorem 4.5 has the following immediate consequence, or corollary.

COROLLARY 4.6 If $G=\langle x\rangle$, then $|G|=o(x)$.

The equality is intended to mean that $|G|$ is infinite iff $o(x)$ is, and that if both sides are finite, then they are equal.

The reader may have noticed that all the examples of cyclic groups that we have looked at are abelian. This is no accident.

THEOREM 4.7 If G is a cyclic group, then G is abelian.

PROOF. Suppose $G=\langle x\rangle$. We have to show that for any elements a,b in G, $ab=ba$. Say $a=x^m$, $b=x^n$, where $m,n\in\mathbb{Z}$. Then

$$ab=(x^m)(x^n)=x^{m+n}=x^{n+m}=x^n x^m=ba,$$

so we are done. $\square$

The converse is false; there exist many noncyclic abelian groups. We have already seen one example: $(\mathbb{Q}, +)$. Another example is **Klein's 4-group**. This group has four elements, e,a,b,c, where the multiplication is given by $ea=ae=a$, $eb=be=b$, $ec=ce=c$, $a^2=b^2=c^2=e$, $ab=ba=c$, $ac=ca=b$, and $bc=cb=a$. Checking associativity for this operation is easy but tedious; it is also a good way to familiarize yourself with the 4-group, so we leave it to you. We can see immediately, however, that we have an identity element and that every element has an inverse—in fact, every element is its own inverse. It is also clear that Klein's 4-group is abelian. However, it is not cyclic. For what could a generator be? Certainly not e; and not a, b, or c, since $\langle a\rangle=\{a,e\}$, $\langle b\rangle=\{b,e\}$, and $\langle c\rangle=\{c,e\}$. (Verify!)

Klein's 4-group is named for the German mathematician Felix Klein (1849–1925). The German word for "4-group" is "Viergruppe," and the 4-group is often denoted by V.

EXERCISES

4.1 Which elements of $(\mathbb{Z}_{10}, \oplus)$ are contained in $\langle 0\rangle$? in $\langle 1\rangle$? $\langle 2\rangle$? $\langle 3\rangle$? $\langle 4\rangle$? $\langle 5\rangle$? $\langle 8\rangle$?

4.2 Let G be the group of all real-valued functions on the real line under addition of functions, and let $f\in G$ be the function such that $f(x)=1$ for all $x\in\mathbb{R}$. Indicate what sort of configuration you would get if you drew the graphs of all the functions in $\langle f\rangle$ on one set of axes.

4.3 Let $X=\{1,2,3,4,5\}$. If A is the element $\{1,4,5\}$ in $(P(X), \triangle)$, how many elements are there in $\langle A\rangle$? What are they?

4.4 In $(\mathbb{Z}_{30}, \oplus)$, find the orders of the elements 3, 4, 6, 7, and 18.

4.5 Let G be a group and let $x \in G$ be an element of order 18. Find the orders of $x^2, x^3, x^4, x^5, x^{12}$.

4.6 List all the elements of $(\mathbb{Z}_{45}, \oplus)$ that are of order 15.

4.7 Let $G = \langle x \rangle$ be a cyclic group of order 24. List all the elements in G that are of order 4.

4.8 The set of even integers forms a group under addition. Is this group cyclic?

4.9 Show that $(\mathbb{Q}^+, \cdot)$ is not cyclic.

4.10 Let $G = \{1, 2, 3, 4, 5, 6\}$ and define an operation $\odot$ on G by $a \odot b = \overline{ab}$, the remainder of ab (mod 7). For instance, $2 \odot 4 = \overline{8} = 1$, and $5 \odot 6 = \overline{30} = 2$.

a) Show that $(G, \odot)$ is a group.

b) Is this group cyclic?

4.11 Let X be a set with at least two elements. Show that $(P(X), \triangle)$ is an abelian noncyclic group.

4.12 Consider the group $(\mathbb{Z}, *)$, where $a * b = a + b - 1$. Is this group cyclic?

4.13 Show that if G is a finite group, then every element of G is of finite order.

4.14 Give an example of an infinite group G such that every element of G has finite order.

4.15 a) Find $(123, 321)$, and find integers x and y such that $123x + 321y = (123, 321)$.

b) Find $(862, 347)$, and find integers x and y such that $862x + 347y = (862, 347)$.

c) Find $(7469, 2464)$, and find integers x and y such that $7469x + 2464y = (7469, 2464)$.

4.16 Prove that if $G = \langle x \rangle$, then $G = \langle x^{-1} \rangle$.

4.17 Prove that if $G = \langle x \rangle$ and G is infinite, then x and x^{-1} are the *only* generators of G.

4.18 Prove parts (ii) and (iii) of Theorem 4.1.

4.19 Prove part (i) of Theorem 4.4.

4.20 Let G be a group and let $a \in G$. An element $b \in G$ is called a *conjugate* of a if there exists an element $x \in G$ such that $b = xax^{-1}$. Show that any conjugate of a has the same order as a.

4.21 Show that for any two elements x, y of any group G, $o(xy) = o(yx)$.

4.22 Let G be an abelian group and let $x, y \in G$. Suppose that x and y are of finite order. Show that xy is of finite order and that, in fact, $o(xy)$ divides $o(x)o(y)$.

4.23 Let G, x, y be as in Exercise 4.22, and assume in addition that $(o(x), o(y)) = 1$. Prove that $o(xy) = o(x)o(y)$.

4.24 Let G be a group and let $x, y \in G$. Assume that $x \neq e$, $o(y) = 2$, and $yxy^{-1} = x^2$. Find $o(x)$.

4.25 Show that if $|G|$ is an even integer, then there is an element $x \in G$ such that $x \neq e$ and $x^2 = e$.

4.26 Let $m, n \in \mathbb{Z}$, not both 0, and let $d \in \mathbb{Z}$. Show that $d = (m, n)$ iff d has the following properties:

  i) d is positive;
  ii) $d|m$ and $d|n$;
  iii) every integer c that divides both m and n divides d.

These three properties are sometimes used to define g.c.d.'s.

4.27 Let p be a prime. Show that if $q_1, \ldots, q_s$ are positive integers and p divides $q_1 q_2 \cdots q_s$, then p divides some q_i.

4.28 Prove the "uniqueness" part of the Fundamental Theorem of Arithmetic (Theorem 0.4). (*Suggestion:* Use the result of Exercise 4.27 and induction on n.)

4.29 Suppose m and n are positive integers and $p_1, \ldots, p_r$ are all the primes that divide m or n or both. Say $m = p_1^{i_1} p_2^{i_2} \cdots p_r^{i_r}$ and $n = p_1^{j_1} p_2^{j_2} \cdots p_r^{j_r}$. Show that $(m, n) = p_1^{k_1} p_2^{k_2} \cdots p_r^{k_r}$, where k_t is the smaller of i_t and j_t, for each t.

4.30 Let $n_1, \ldots, n_k$ be integers, not all 0. The *greatest common divisor* of $n_1, \ldots, n_k$, denoted by $(n_1, \ldots, n_k)$, is the largest integer that divides all of $n_1, n_2, \ldots, n_k$. Show that there exist integers $a_1, \ldots, a_k$ such that

$$a_1 n_1 + a_2 n_2 + \cdots + a_k n_k = (n_1, \ldots, n_k).$$

[*Suggestion:* Use induction on k. Use the inductive hypothesis to show that

$$(n_1, \ldots, n_k) = ((n_1, \ldots, n_{k-1}), n_k),$$

and apply Theorem 4.2.]

4.31 If m and n are integers, we define their **least common multiple**, $[m, n]$, as follows. If $m = 0$ or $n = 0$, we set $[m, n] = 0$; otherwise we let $[m, n]$ be the smallest positive integer that is divisible by both m and n.

  a) Show that if m and n are both positive and $m = p_1^{i_1} p_2^{i_2} \cdots p_r^{i_r}$, $n = p_1^{j_1} p_2^{j_2} \cdots p_r^{j_r}$, as in Exercise 4.29, then $[m, n] = p_1^{l_1} p_2^{l_2} \cdots p_r^{l_r}$, where l_t is the larger of i_t and j_t, for each t.

  b) Show that if m and n are both positive, then

$$mn = (m, n)[m, n].$$

4.32 Let G be an abelian group, and let x and y be elements of G such that $o(x) = m$ and $o(y) = n$. Show that G has an element z such that $o(z)$ is the least common multiple of m and n.

SUBGROUPS

Up to this point we have been considering groups as separate entities, unrelated to each other. Even so, you have probably observed that some groups sit inside others. For example, in $(\mathbb{Z}, +)$, the set $2\mathbb{Z}$ of even integers is itself a group under $+$: Addition is a binary operation on $2\mathbb{Z}$ since the sum of two even integers is even; addition is associative on $2\mathbb{Z}$ since it is associative on all of $\mathbb{Z}$; $2\mathbb{Z}$ contains the identity element 0 of $(\mathbb{Z}, +)$; and if $x \in 2\mathbb{Z}$ then $-x \in 2\mathbb{Z}$, so $2\mathbb{Z}$ contains the inverse of each of its elements.

Indeed, one of the most natural questions one can ask about a group G is "What groups sit inside G?" Those that do are called *subgroups* of G.

DEFINITION A subset H of a group $(G, *)$ is called a **subgroup** of G if the elements of H form a group under $*$.

It is worth emphasizing the "under $*$." For example, $(\mathbb{Q}^+, \cdot)$ is a group and $(\mathbb{Q}, +)$ is a group, and $\mathbb{Q}^+ \subseteq \mathbb{Q}$, but $(\mathbb{Q}^+, \cdot)$ is *not* a subgroup of $(\mathbb{Q}, +)$ because the operation on $(\mathbb{Q}^+, \cdot)$ is not the operation on $(\mathbb{Q}, +)$.

Observe that if H is a subgroup of G, then H cannot be empty because H must contain an identity element. In fact, the identity element of H must be e, the identity element of G. For suppose e' is the identity element of H; then in particular $e' * e' = e'$, a relationship between elements of the group G. Thus, multiplying by $(e')^{-1}$ in G, we get $e' = e$.

It is convenient to have a more compact criterion for a subset of a group to constitute a subgroup.

THEOREM 5.1 Let H be a nonempty subset of a group G. Then H is a subgroup of G if and only if the following two conditions are satisfied:

i) for all $a, b \in H$, $ab \in H$, and

ii) for all $a \in H$, $a^{-1} \in H$.

Condition (i) is expressed by saying that H is **closed** under the operation in G, and condition (ii) is expressed by saying that H is **closed under inverses**.

PROOF OF THE THEOREM. If H is, in fact, a subgroup of G then it is clear that (i) is satisfied. As for (ii), if we let a_{-1} denote the inverse of a in H, then $aa_{-1} = e$ (we have remarked that the identity element of H is the same as the identity element of G) in G, which implies that a_{-1} is in fact a^{-1}, the inverse of a in G. Thus $a^{-1} \in H$ for any $a \in H$, so (ii) is satisfied.

Conversely, assume that the nonempty subset H is closed under the operation $*$ in G and that H is closed under inverses. To show that H is then a group under $*$ it suffices to check that $e \in H$ and that associativity holds in H. Since H is nonempty by assumption, we can let x denote some element of H. Then by (ii) $x^{-1} \in H$, so by (i) $xx^{-1} \in H$. But $xx^{-1} = e$, so $e \in H$. Finally, H inherits associativity from G: if $a,b,c \in H$ then $a,b,c \in G$, so $(ab)c = a(bc)$ by associativity in G. $\square$

Examples

1. $(\mathbb{Q}^+, \cdot)$ is a subgroup of $(\mathbb{R}^+, \cdot)$, which is, in turn, a subgroup of $(\mathbb{R} - \{0\}, \cdot)$.

2. Let G be any group and let $a \in G$.. Then $\langle a \rangle$ is a subgroup of G. For clearly $\langle a \rangle \neq \varnothing$, and $\langle a \rangle$ is closed under multiplication since if $a^j, a^k \in \langle a \rangle$ then $a^j a^k = a^{j+k} \in \langle a \rangle$. Finally, $\langle a \rangle$ is closed under inverses since if $a^j \in \langle a \rangle$ then $(a^j)^{-1} = a^{-j} \in \langle a \rangle$.

As a specific example, if $G = (\mathbb{Z}, +)$ and $a = 2$, then $\langle a \rangle$ is the subgroup $(2\mathbb{Z}, +)$.

3. Let's find all the subgroups of Klein's 4-group, $V = \{e, a, b, c\}$ with $a^2 = b^2 = c^2 = e$, $ab = ba = c$, $ac = ca = b$, $bc = cb = a$. The only subgroup that contains none of a, b, or c is obviously $\{e\}$. If a subgroup contains just one of a, b, or c then it is either $\langle a \rangle$, $\langle b \rangle$, or $\langle c \rangle$. If it contains two of a,b,c then by the definition of multiplication in V, it contains the third as well, and since it contains e it must then be V. Thus the subgroups are $\langle e \rangle$, $\langle a \rangle$, $\langle b \rangle$, $\langle c \rangle$, and V.

Klein's 4-group is thus an example of an abelian noncyclic group with the property that all of its proper subgroups are cyclic. (A subgroup H of G is **proper** if $H \neq G$.)

We sometimes make a schematic picture of the subgroups of a group by drawing what is called a **subgroup lattice**. The subgroup lattice for Klein's

4-group is

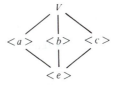

A line going upward from one group to another indicates that the bottom group is a subgroup of the top one.

4. If $G = (\mathbb{Z}, +)$ and n is any integer, then the set $n\mathbb{Z} = \langle n \rangle$ of multiples of n forms a subgroup of G. In particular for $n = 0$ we get the subgroup $\{0\}$ consisting of just the identity element, and for $n = 1$ we get G itself.

In fact the $n\mathbb{Z}$ are *all* the subgroups of $(\mathbb{Z}, +)$. For if H is a subgroup other than $\{0\}$ there exist positive integers in H. (Why?) If we let n be the smallest positive integer in H, then we claim that H is $n\mathbb{Z}$. Clearly, $n\mathbb{Z} \subseteq H$ since $n \in H$ and H is a subgroup. But also $H \subseteq n\mathbb{Z}$, for if $h \in H$ we can write $h = qn + r$, with $0 \leqslant r < n$. Then

$$r = h - qn \in H \quad (\text{why?}),$$

which contradicts the minimality of n *unless* $r = 0$. Thus r is 0, so $h = qn$ and any $h \in H$ is a multiple of n.

5. Let n be a positive integer and consider $(\mathbb{Z}_n, \oplus)$. We have the subgroups

$$\langle 0 \rangle, \langle 1 \rangle, \langle 2 \rangle, \ldots, \langle n-1 \rangle$$

(not all of which are distinct if $n > 2$: for example, $\langle 1 \rangle = \langle n-1 \rangle$ since $n-1$ is the inverse of 1), and an argument like that in Example 4 would show that these are, in fact, all the subgroups of $(\mathbb{Z}_n, \oplus)$. However, rather than go through the argument in the context of $(\mathbb{Z}_n, \oplus)$, we will prove a general result that covers Example 4 and $(\mathbb{Z}_n, \oplus)$ simultaneously.

THEOREM 5.2 Let G be a cyclic group. Then every subgroup of G is cyclic.

PROOF. Suppose $G = \langle x \rangle$, and let H be a subgroup of G. If $H = \{e\}$ then clearly H is cyclic, so assume $H \neq \{e\}$. Let n be the smallest positive integer such that $x^n \in H$. (Why does n exist?) We assert that $H = \langle x^n \rangle$. For if $h \in H$ then $h = x^m$ for some integer m, because $h \in G = \langle x \rangle$. Write $m = qn + r$, where $0 \leqslant r < n$. Then

$$h = x^{qn+r} = x^{qn}x^r = (x^n)^q x^r, \quad \text{and} \quad (x^n)^q \in H.$$

So $x^r = x^{-nq}h \in H$. Since $0 \leqslant r < n$, the definition of n implies that $r = 0$, so $m = qn$ and

$$h = x^m = (x^n)^q \in \langle x^n \rangle.$$

Thus $H \subseteq \langle x^n \rangle$, and since it is obvious that $\langle x^n \rangle \subseteq H$, the proof is complete. ☐

It is often true in mathematics that the *proof* of a theorem contains substantially more information than the statement of the theorem itself. Theorem 5.2 is a case in point, and you are urged to study the proof carefully. For example, the proof tells us not only that every subgroup is cyclic, but also how to find a generator for each one.

There are, of course, questions about the subgroups of $(\mathbb{Z}_n, \oplus)$ that are left unanswered by this theorem. Which of $\langle 0 \rangle, \langle 1 \rangle, \langle 2 \rangle, \ldots, \langle n-1 \rangle$ coincide? How many distinct subgroups does $(\mathbb{Z}_n, \oplus)$ have? We shall return to these questions after we consider some more examples of subgroups.

Examples (continued)

6. Let G be the group of real-valued functions on the real line, under addition of functions, and let H be the set of continuous functions in G. Then H is a subgroup of G since the sum of two continuous functions is continuous (closure of H under addition of functions) and if f is continuous, then $-f$ is continuous (closure of H under inverses).

7. Let $G = GL(2, \mathbb{R})$. Let H be the subset of G consisting of all matrices of the form $\begin{pmatrix} a & b \\ 0 & d \end{pmatrix}$, where $ad \neq 0$. Then H is a subgroup of G. To see that H is closed under matrix multiplication, observe that

$$\begin{pmatrix} a & b \\ 0 & d \end{pmatrix}\begin{pmatrix} e & f \\ 0 & h \end{pmatrix} = \begin{pmatrix} ae & af + bh \\ 0 & dh \end{pmatrix},$$

and if $ad \neq 0$ and $eh \neq 0$ then $(ae)(dh) = (ad)(eh) \neq 0$. To check closure under inverses, observe that if $ad \neq 0$, then the inverse of $\begin{pmatrix} a & b \\ 0 & d \end{pmatrix}$ is

$$\begin{pmatrix} 1/a & -b/ad \\ 0 & 1/d \end{pmatrix},$$

and $(1/a) \cdot (1/d) \neq 0$.

8. If G is any group, then the **center** of G, denoted by $Z(G)$, is the subset consisting of those elements that commute with everything in G. Thus

$$Z(G) = \{z \in G \mid zx = xz \quad \text{for all } x \in G\}.$$

(The Z comes from the German word "Zentrum.")

Of course, if G is abelian, then $Z(G) = G$; but if G is nonabelian, then $Z(G) \subsetneq G$. As a specific example, let's try to figure out what the center of $GL(2, \mathbb{R})$ looks like.

Suppose $\begin{pmatrix} a & b \\ c & d \end{pmatrix}$ commutes with everything in $GL(2, \mathbb{R})$. Then since $\begin{pmatrix} 0 & 1 \\ 1 & 0 \end{pmatrix} \in GL(2, \mathbb{R})$, we have

$$\begin{pmatrix} a & b \\ c & d \end{pmatrix}\begin{pmatrix} 0 & 1 \\ 1 & 0 \end{pmatrix} = \begin{pmatrix} 0 & 1 \\ 1 & 0 \end{pmatrix}\begin{pmatrix} a & b \\ c & d \end{pmatrix},$$

that is,

$$\begin{pmatrix} b & a \\ d & c \end{pmatrix} = \begin{pmatrix} c & d \\ a & b \end{pmatrix}.$$

This means that $a = d$ and $b = c$, so

$$\begin{pmatrix} a & b \\ c & d \end{pmatrix} = \begin{pmatrix} a & b \\ b & a \end{pmatrix}.$$

Furthermore, since $\begin{pmatrix} 1 & 1 \\ 0 & 1 \end{pmatrix} \in GL(2, \mathbb{R})$, we also have

$$\begin{pmatrix} a & b \\ b & a \end{pmatrix}\begin{pmatrix} 1 & 1 \\ 0 & 1 \end{pmatrix} = \begin{pmatrix} 1 & 1 \\ 0 & 1 \end{pmatrix}\begin{pmatrix} a & b \\ b & a \end{pmatrix},$$

that is,

$$\begin{pmatrix} a & a+b \\ b & b+a \end{pmatrix} = \begin{pmatrix} a+b & b+a \\ b & a \end{pmatrix}.$$

Thus $b = 0$, so every element of the center has the form $\begin{pmatrix} a & 0 \\ 0 & a \end{pmatrix}$, $a \neq 0$. It is easy to see that, conversely, all elements of this form are in the center, so

$$Z(GL(2, \mathbb{R})) = \left\{ \begin{pmatrix} a & 0 \\ 0 & a \end{pmatrix} \Big| a \neq 0 \right\}.$$

In this case the center is clearly a subgroup, and it is not hard to see that the center is *always* a subgroup, no matter what group G is (Exercise 5.22).

9. Let $G = GL(2, \mathbb{C})$, the general linear group of degree two over the complex numbers. This is the set of all invertible matrices

$$\begin{pmatrix} a_1 + a_2 i & b_1 + b_2 i \\ c_1 + c_2 i & d_1 + d_2 i \end{pmatrix},$$

where the a's, b's, c's, and d's are real numbers and $i^2 = -1$, under matrix multiplication. [All you need to know about multiplying complex numbers at this point is that

$$(a + bi)(c + di) = (ac - bd) + (bc + ad)i.$$

That is, you just multiply it out and replace i^2 by -1.]

We are going to consider a subgroup of G that is of order 8, called the **group of unit quaternions**.

Let

$$J = \begin{pmatrix} i & 0 \\ 0 & -i \end{pmatrix}, \quad K = \begin{pmatrix} 0 & 1 \\ -1 & 0 \end{pmatrix}, \quad \text{and} \quad L = JK = \begin{pmatrix} 0 & i \\ i & 0 \end{pmatrix}.$$

Let $Q_8 = \{I, -I, J, -J, K, -K, L, -L\}$, where I (for "identity") is, of course, $\begin{pmatrix} 1 & 0 \\ 0 & 1 \end{pmatrix}$, and $-J$ is, for example, $\begin{pmatrix} -i & 0 \\ 0 & i \end{pmatrix}$. To show that Q_8 is in fact a subgroup of $GL(2, \mathbb{C})$ it suffices to check that Q_8 is closed under matrix multiplication and inverses. Closure under multiplication follows from the facts that

$$J^2 = K^2 = L^2 = -I$$

and $KJ = -L$, $JK = L$, $KL = J$, $LK = -J$, $JL = -K$, $LJ = K$.

Closure under inverses follows from the fact that $J^{-1} = -J$, $K^{-1} = -K$, and $L^{-1} = -L$. [For example, $J(-J) = -J^2 = -(-I) = I$.] However, it turns out that because Q_8 is a *finite* subset of $GL(2, \mathbb{C})$ it was really enough to check just closure under multiplication.

Thus Q_8 provides a nice illustration of the following elegant little theorem.

THEOREM 5.3 Let G be a group and let H be a *finite* nonempty subset of G. Then if H is closed under multiplication in G, H is a subgroup of G.

PROOF. All we have to do is to use the *finiteness* of H to show that H is also closed under inverses. Let $h \in H$; we have to find $x \in H$ such that $hx = e$. What do we know is in H? Well, H is closed under multiplication, so certainly $h, h^2, h^3, h^4, \ldots$ are all in H. But H is finite, so these elements can't all be distinct. Say $h^n = h^m$, where $n > m$. Then:

$$h^{n-m} = e, \quad \text{so} \qquad \qquad [5.1]$$

$$h(h^{n-m-1}) = e. \qquad \qquad [5.2]$$

Since $n - m - 1 \geqslant 0$, we have either (1) $n - m - 1 = 0$, in which case $n - m = 1$, so Eq. [5.1] implies that $h = e$, and we can take $x = e \in H$, or (2) $n - m - 1 > 0$, in which case $h^{n-m-1} \in H$ and Eq. [5.2] implies that we can take $x = h^{n-m-1}$. □

In Exercise 5.17 you are asked to consider the question of whether it would have been enough to assume only that H was closed under inverses.

The group of unit quaternions will be useful to us as an example when we discuss what are called *normal* subgroups. For now we note that the multiplication in Q_8 can be remembered by using the diagram

Going around the diagram clockwise we read off products: $JK = L$, $LJ = K$, $KL = J$. Going around counterclockwise we attach minus signs: $JL = -K$, and so on. The group Q_8 can be defined abstractly as the set of eight symbols $\{I, -I, J, -J, K, -K, L, -L\}$ with multiplication defined by $J^2 = K^2 = L^2 = -I$, $JK = L$, $KL = J$, $LJ = K$, $JL = -K$, $LK = -J$, $KJ = -L$, and I the identity element. If we take this approach then the fact that Q_8 satisfies the group axioms must be checked from scratch, and verifying associativity becomes rather tedious. This is why we have chosen to introduce Q_8 as a subgroup of the known group $GL(2, \mathbb{C})$ instead.

One cannot think for very long about subgroups without running into the following questions. Suppose H and K are subgroups of G. When is $H \cap K$ a subgroup of G? How about $H \cup K$?

Let's consider these questions in the context of $(\mathbb{Z}, +)$ for a moment. Recall that the subgroups of $(\mathbb{Z}, +)$ are just the sets $n\mathbb{Z}$, where n is an integer. What is $2\mathbb{Z} \cap 3\mathbb{Z}$? Since an integer is divisible by both 2 and 3 if and only if it is divisible by 6, $2\mathbb{Z} \cap 3\mathbb{Z} = 6\mathbb{Z}$. More generally,

$$m\mathbb{Z} \cap n\mathbb{Z} = k\mathbb{Z},$$

where k is the least common multiple of m and n. So in this case, at least, the intersection of two subgroups is always a subgroup.

How about unions? Well, $2\mathbb{Z} \cup 3\mathbb{Z}$ is *not* a subgroup of $(\mathbb{Z}, +)$. For example, $2 + 3 = 5 \notin 2\mathbb{Z} \cup 3\mathbb{Z}$, so $2\mathbb{Z} \cup 3\mathbb{Z}$ is not closed under addition. However, $2\mathbb{Z} \cup 6\mathbb{Z}$ *is* a subgroup, since $2\mathbb{Z} \cup 6\mathbb{Z} = 2\mathbb{Z}$. (Why?) After trying a few more examples, one would get the idea that the union of two subgroups of $(\mathbb{Z}, +)$ is a subgroup if and only if one of the two subgroups is contained in the other. This idea would be correct, and here is the general result.

THEOREM 5.4 Let H and K be subgroups of a group G. Then:

i) $H \cap K$ is always a subgroup of G; and

ii) $H \cup K$ is a subgroup if and only if one of H, K is contained in the other.

Another way of stating (ii) is that $H \cup K$ is a subgroup if and only if $H \cup K$ is either H or K.

PROOF OF THE THEOREM.

The proof of part (i) is left as an exercise.

ii) Clearly if $H \cup K$ is either H or K, then $H \cup K$ is a subgroup of G. Conversely, let us suppose that $H \not\subseteq K$ and $K \not\subseteq H$, and try to show that $H \cup K$ is not a subgroup of G. What we know is that there is some $h \in H$ such that $h \notin K$, and there is some $k \in K$ such that $k \notin H$. We show that $hk \notin H \cup K$, so that $H \cup K$ is not closed under the operation in G. (In the above example, H was $2\mathbb{Z}$, K was $3\mathbb{Z}$, h was 2, k was 3, and $2 + 3 = 5 \notin H \cup K$.)

Could hk be in H? If it were then, since $h^{-1} \in H$, we would have $h^{-1}(hk) \in H$ (why?), that is, $k \in H$. But $k \notin H$. Similarly, hk is not in K either, so $hk \notin K \cup H$. $\square$

Observe that $H \cup K$ will always be closed under inverses, so whenever it fails to be a subgroup, the reason will always be that it is not closed under the group operation.

We conclude this section by discussing the subgroups of a cyclic group G of order n. We already know that all the subgroups of G are cyclic.

THEOREM 5.5 Let $G = \langle x \rangle$ be a cyclic group of order n. Then:

i) For any positive integer m, G has a subgroup of order m if and only if m divides n.

ii) If in fact m does divide n, then G has a *unique* subgroup of order m.

iii) Two powers x^r and x^s of x generate the same subgroup of G if and only if $(r, n) = (s, n)$.

PROOF. i) First suppose that m divides n. Then $\langle x^{n/m} \rangle$ is a subgroup of G. By Corollary 4.6, x has order n, so by Theorem 4.4, $x^{n/m}$ has order

$$\frac{n}{(n/m, n)} = \frac{n}{(n/m)} = m.$$

Thus by Corollary 4.6 again, $\langle x^{n/m} \rangle$ is a subgroup of G of order m.

Conversely, assume that G has a subgroup H of order m. H must be cyclic, say $H = \langle x^k \rangle$. Then

$$|H| = o(x^k) = n/(k, n),$$

so

$$m = n/(k, n).$$

This yields $m(k, n) = n$, and therefore m divides n.

ii) Let H_1 and H_2 be two subgroups of order m, where m divides n. Then, as in the proof of Theorem 5.2, we have $H_1 = \langle x^{k_1} \rangle$ and $H_2 = \langle x^{k_2} \rangle$, where x^{k_1} and x^{k_2} are the smallest positive powers of x that lie in H_1 and H_2, respectively. (In the present case this is true even if H_1 or H_2 is $\{e\}$, because x has finite order.) Our task is to show that $k_1 = k_2$, for this would imply that $H_1 = H_2$, thus finishing the proof. Now

$$n/(k_1, n) = |H_1| = m = |H_2| = n/(k_2, n),$$

so $(k_1, n) = (k_2, n)$. How can we get from this to $k_1 = k_2$?

We will show that $(k_1, n) = k_1$ and $(k_2, n) = k_2$. All we have to do to show that $(k_1, n) = k_1$ is to show that k_1 divides n. But if t is *any* integer such that $x^t \in H_1$, then k_1 divides t (see the proof of Theorem 5.2). Since $x^n = e \in H_1$, k_1 divides n.

We have shown that $(k_1, n) = k_1$, and the same argument shows that $(k_2, n) = k_2$. Thus $k_1 = k_2$ and the proof is complete.

iii) Consider two elements x^r, x^s of G. By part (ii), we see that

$$\langle x^r \rangle = \langle x^s \rangle \quad \text{iff}$$
$$|\langle x^r \rangle| = |\langle x^s \rangle| \quad \text{iff}$$
$$o(x^r) = o(x^s) \quad \text{iff}$$
$$\frac{n}{(r,n)} = \frac{n}{(s,n)} \quad \text{iff}$$
$$(r,n) = (s,n). \quad \square$$

Parts (i) and (ii) of the theorem imply that a cyclic group of order n has exactly $d(n)$ subgroups, where $d(n)$ denotes the number of positive divisors of n. The function $d(n)$ [sometimes denoted by $\tau(n)$] will be familiar to you if you have studied number theory.

COROLLARY 5.6 If $G = \langle x \rangle$ is cyclic of order n and $d_1, d_2, \ldots, d_{d(n)}$ are the distinct positive divisors of n, then

$$\langle x^{d_1} \rangle, \langle x^{d_2} \rangle, \ldots, \langle x^{d_{d(n)}} \rangle$$

are the distinct subgroups of G.

PROOF. Observe that, since each d_i divides n, we have $(d_i, n) = d_i$. Hence if $d_i \neq d_j$ then $(d_i, n) \neq (d_j, n)$, and therefore $\langle x^{d_i} \rangle \neq \langle x^{d_j} \rangle$ by part (iii) of Theorem 5.5. Thus all the indicated subgroups are distinct, and since there are $d(n)$ of them, they must give all $d(n)$ subgroups of G. $\square$

Example Let $G = (\mathbb{Z}_{12}, \oplus)$. Then $G = \langle 1 \rangle$ and $|G| = 12$. Since the positive divisors of 12 are 1, 2, 3, 4, 6, and 12, the distinct subgroups of G are generated by the 1st, 2nd, 3rd, 4th, 6th, and 12th powers of 1, namely $1, 2 \cdot 1$, $3 \cdot 1, 4 \cdot 1, 6 \cdot 1$, and $12 \cdot 1$. Thus the subgroups of G are $\langle 1 \rangle$, $\langle 2 \rangle$, $\langle 3 \rangle$, $\langle 4 \rangle$, $\langle 6 \rangle$, and $\langle 0 \rangle$.

Observe that $\langle 5 \rangle = \langle 1 \rangle = G$ since $(5, 12) = (1, 12)$; similarly $\langle 7 \rangle = \langle 1 \rangle$ and $\langle 11 \rangle = 1$. We have $\langle 8 \rangle = \langle 4 \rangle$ since $(8, 12) = (4, 12)$; $\langle 9 \rangle = \langle 3 \rangle$; $\langle 10 \rangle = \langle 2 \rangle$.

If we arrange the subgroups of $(\mathbb{Z}_{12}, \oplus)$ in a lattice, it comes out like this:

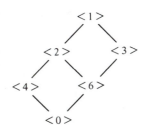

Here is the situation for infinite cyclic groups.

THEOREM 5.7 Let $G=\langle x \rangle$ be an infinite cyclic group. Then $\langle e \rangle, \langle x \rangle, \langle x^2 \rangle,$ $\langle x^3 \rangle, \langle x^4 \rangle, \ldots$ are all the distinct subgroups of G.

PROOF. This is a good exercise. $\square$

EXERCISES

5.1 In each case, determine whether or not H is a subgroup of G.

a) $G=(\mathbb{R},+)$; $H=\mathbb{Q}$

b) $G=(\mathbb{Q},+)$; $H=\mathbb{Z}$

c) $G=(\mathbb{Z},+)$; $H=\mathbb{Z}^+$

d) $G=(\mathbb{Q}-\{0\},\cdot)$; $H=\mathbb{Q}^+$

e) $G=(\mathbb{Z}_8,\oplus)$; $H=\{0,2,4\}$

f) $G=$ the set of 2-tuples of real numbers (a,b) under addition of 2-tuples; $H=$ the subset consisting of all 2-tuples such that $b=-a$

g) $G=Q_8$; $H=\{I,J,K\}$

h) $G=(P(X),\triangle)$; $H=\{\varnothing, A, B, A\triangle B\}$, where A, B are two elements of G

i) $G=(P(X),\triangle)$; $H=P(Y)$, where $Y\subseteq X$.

5.2 Let G be the group of real-valued functions on the real line, under addition of functions. Let H be the set of differentiable functions in G. Show that H is a subgroup of G.

5.3 Let H be the set of elements $\begin{pmatrix} a & b \\ c & d \end{pmatrix}$ of $GL(2,\mathbb{R})$ such that $ad-bc=1$. Show that H is a subgroup of $GL(2,\mathbb{R})$. H is called the **special linear group of degree 2 over** $\mathbb{R}$ and is denoted by $SL(2,\mathbb{R})$.

5.4 a) How many subgroups does $(\mathbb{Z}_{18},\oplus)$ have? What are they?

b) How many subgroups does $(\mathbb{Z}_{35},\oplus)$ have? What are they?

c) How many subgroups does $(\mathbb{Z}_{36},\oplus)$ have? What are they?

5.5 Find all the subgroups of Q_8. Show that Q_8 is an example of a nonabelian group with the property that all its proper subgroups are cyclic.

5.6 a) Let G be a cyclic group of order n. Show that if m is a positive integer, then G has an element of order m iff m divides n.

b) Let G be a cyclic group of order 40. List all the possibilities for the orders of elements of G.

5.7 Let $G=\langle x \rangle$ be a cyclic group of order n. Show that x^m is a generator of G if and only if $(m,n)=1$. Thus the number of generators of a cyclic group of order n is the number of integers m in the set $\{0,1,\ldots,n-1\}$ such that $(m,n)=1$. This number is denoted by $\phi(n)$; ϕ is called *Euler's function* and plays a prominent role in number theory.

5.8 Let $G = \langle x \rangle$ be a cyclic group of order 144. How many elements are there in the subgroup $\langle x^{26} \rangle$?

5.9 Let $m\mathbb{Z}$ and $n\mathbb{Z}$ be subgroups of $(\mathbb{Z}, +)$. What condition on m and n is equivalent to $m\mathbb{Z} \subseteq n\mathbb{Z}$? What condition on m and n is equivalent to $m\mathbb{Z} \cup n\mathbb{Z}$ being a subgroup of $(\mathbb{Z}, +)$?

5.10 Prove that every subgroup of an abelian group is abelian.

5.11 Let G be an abelian group, and let n be a positive integer. Let H be the subset of G consisting of all $x \in G$ such that $x^n = e$. Show that H is a subgroup of G.

5.12 Find the center of:

a) V;

b) Q_8.

5.13 Let H be the group introduced in Example 7 on p. 46. Find $Z(H)$.

5.14 Prove that the intersection of two subgroups of a group G is itself a subgroup of G.

5.15 Show that if H and K are subgroups of the group G, then $H \cup K$ is closed under inverses.

5.16 Give an example of a group G and a subset H of G such that H is closed under multiplication but H is *not* a subgroup of G.

5.17 Suppose H is a nonempty finite subset of a group G and H is closed under inverses. Must H be a subgroup of G? Either prove that it must, or give a counterexample.

5.18 a) Show that it is impossible for a group G to be the union of two proper subgroups.

b) Give an example of a group that is the union of three proper subgroups.

5.19 Let $G = \langle x \rangle$ be an infinite cyclic group. Show that all the distinct subgroups of G are $\langle e \rangle, \langle x \rangle, \langle x^2 \rangle, \langle x^3 \rangle, \langle x^4 \rangle, \langle x^5 \rangle, \dots$.

5.20 Let G be a finite group with no subgroups other than $\{e\}$ or G itself. Prove that G is either the trivial group $\{e\}$ or a cyclic group of prime order.

5.21 Let $G = \langle x \rangle$ be a cyclic group of order n. Find a condition on the integers r and s that is equivalent to $\langle x^r \rangle \subseteq \langle x^s \rangle$.

5.22 Let G be a group. Prove that $Z(G)$ is a subgroup of G.

5.23 Let G be a group, and let $g \in G$. Define the centralizer, $Z(g)$, of g in G to be the subset

$$Z(g) = \{x \in G \mid xg = gx\}.$$

Prove that $Z(g)$ is a subgroup of G.

5.24 Let G be a group and let H be a nonempty subset of G such that whenever $x, y \in H$ we have $xy^{-1} \in H$. Prove that H is a subgroup of G.

5.25 Let G be a group and let a be some fixed element of G. Let H be a subgroup of G and let aHa^{-1} be the subset of G consisting of all elements that are of the form aha^{-1}, with $h \in H$. Show that aHa^{-1} is a subgroup of G. It is called the *conjugate subgroup of H by a*.

5.26 Let H be a subgroup of the group G and let $N(H) = \{a \in G \mid aHa^{-1} = H\}$. (See Exercise 5.25 for the definition of aHa^{-1}.) Prove that $N(H)$ is a subgroup of G.

5.27 Let G be a finite abelian group. Show that G is cyclic iff G has the property that for every positive integer n, there are at most n elements x in G such that $x^n = e$.

DIRECT PRODUCTS

In dealing with an abstract concept, it is very useful to have a good supply of concrete examples on hand. The examples make the abstraction come to life, and they also provide us with a means of testing out general ideas on specific cases. In the preceding section, we saw that new examples of groups can sometimes be found sitting inside old ones. Now we take the opposite tack by considering how we can "patch together" given groups to make new ones.

Actually there are a number of different ways of doing this. We will consider the simplest and most frequently used method, called the *direct product* construction.

Suppose G and H are groups (not necessarily distinct). To form the direct product of G and H, we consider the set of all ordered pairs (g,h), where $g \in G$ and $h \in H$. We introduce an operation on this set by multiplying componentwise:

$$(g_1, h_1)(g_2, h_2) = (g_1 g_2, h_1 h_2),$$

where $g_1 g_2$ is computed in G and $h_1 h_2$ is computed in H. It is clear that this definition gives us a binary operation, and we have associativity as a consequence of associativity in G and H:

$$\left[(g_1, h_1)(g_2, h_2) \right] (g_3, h_3) = (g_1 g_2, h_1 h_2)(g_3, h_3) = \left[(g_1 g_2)g_3, (h_1 h_2)h_3 \right]$$
$$= \left[g_1(g_2 g_3), h_1(h_2 h_3) \right] = (g_1, h_1)(g_2 g_3, h_2 h_3) = (g_1, h_1)\left[(g_2, h_2)(g_3, h_3) \right].$$

The identity element is (e_G, e_H), where e_G and e_H are the respective identity elements of G and H, and the inverse of (g,h) is (g^{-1}, h^{-1}).

Thus we have a new group, which we denote by $G \times H$ and call the **direct product** of G and H. The groups G and H are called **factors** of the product.

In a completely analogous fashion, we can form the direct product

$$G_1 \times G_2 \times \cdots \times G_n$$

of n groups. The elements of this group are n-tuples $(g_1, g_2, \ldots, g_n)$ with $g_i \in G_i$, and the multiplication is defined componentwise. As a matter of fact, there is no reason why we have to restrict ourselves to finitely many factors, but we will rarely use infinitely many.

Examples

1. Let $G_1 = G_2 = \cdots = G_n = (\mathbb{R}, +)$. Then

$$G_1 \times G_2 \times \cdots \times G_n = \mathbb{R} \times \mathbb{R} \times \cdots \times \mathbb{R}$$

is ordinary n-space $\mathbb{R}^n$ under addition of n-tuples.

2. Consider $\mathbb{Z}_2 \times \mathbb{Z}_2$, where the operation on each factor is addition mod 2. This is a group of order 4, and already reveals some interesting things about direct products.

First of all, $\mathbb{Z}_2 \times \mathbb{Z}_2$ is not cyclic, although both factors are cyclic. Denoting the operation on $\mathbb{Z}_2 \times \mathbb{Z}_2$ by $+$, for simplicity, we have

$$(0, 1) + (0, 1) = (0, 0),$$
$$(1, 0) + (1, 0) = (0, 0),$$
and
$$(1, 1) + (1, 1) = (0, 0),$$

so that every nonidentity element has order 2, and there is no element of order 4 to generate the group.

Notice something else. In general, a direct product $G \times H$ has certain "obvious" subgroups, because if A is a subgroup of G and B is a subgroup of H, then $A \times B$ is a subgroup of $G \times H$ (Exercise 6.5). $\mathbb{Z}_2 \times \mathbb{Z}_2$ points out that $G \times H$ may have subgroups *other* than those of the form $A \times B$. For instance, if the cyclic subgroup

$$\langle (1, 1) \rangle = \{ (1, 1), (0, 0) \}$$

were of this form, then A and B would both have to be $\mathbb{Z}_2$; but then $A \times B$ would be the whole group $\mathbb{Z}_2 \times \mathbb{Z}_2$, not $\langle (1, 1) \rangle$.

3. For $\mathbb{Z}_2 \times \mathbb{Z}_3$, things are different. The group has order 6, and *is* cyclic, because $(1, 1)$ has order 6 [its first six powers are $(1, 1)$, $(0, 2)$, $(1, 0)$, $(0, 1)$, $(1, 2)$, $(0, 0)$]. Also, by Theorem 5.5, we know that there is a unique subgroup of order m, for each m dividing 6. Since

$$\langle 0 \rangle \times \langle 0 \rangle, \quad \mathbb{Z}_2 \times \langle 0 \rangle, \quad \langle 0 \rangle \times \mathbb{Z}_3, \quad \text{and} \quad \mathbb{Z}_2 \times \mathbb{Z}_3$$

are subgroups of order 1, 2, 3, and 6, these are the only subgroups, so every subgroup has the form $A \times B$.

The following result goes a long way toward explaining the difference between Examples 2 and 3.

THEOREM 6.1 Let $G = G_1 \times G_2 \times \cdots \times G_n$.

i) If $g_i \in G_i$ for $1 \leqslant i \leqslant n$, and each g_i has finite order, then $o((g_1, g_2, \ldots, g_n))$ is the least common multiple of $o(g_1), o(g_2), \ldots, o(g_n)$.

ii) If each G_i is a cyclic group of finite order, then G is cyclic iff $|G_i|$ and $|G_j|$ are relatively prime for $i \neq j$.

PROOF. i) If m is a positive integer, then

$$(g_1, g_2, \ldots, g_n)^m = (g_1^m, g_2^m, \ldots, g_n^m).$$

It follows, by Theorem 4.4, that $(g_1, g_2, \ldots, g_n)^m = (e_{G_1}, e_{G_2}, \ldots, e_{G_n})$ iff m is divisible by each $o(g_i)$. Thus $o((g_1, g_2, \ldots, g_n))$ is the smallest positive integer that is divisible by each $o(g_i)$.

ii) If G is cyclic, let $g = (g_1, g_2, \ldots, g_n)$ be a generator. Then for $1 \leqslant i \leqslant n$, g_i generates G_i (why?), so by Corollary 4.6 we have $o(g_i) = |G_i|$. Thus, by part (i), $o(g)$ is the least common multiple of $|G_1|, |G_2|, \ldots, |G_n|$. But since g generates G,

$$o(g) = |G| = |G_1| \cdot |G_2| \cdot \cdots \cdot |G_n|$$

(see Exercise 6.4). We conclude that the least common multiple of $|G_1|, |G_2|, \ldots, |G_n|$ is $|G_1| \cdot |G_2| \cdot \cdots \cdot |G_n|$, and this means that $|G_i|$ and $|G_j|$ are relatively prime if $i \neq j$.

Conversely, if $|G_i|$ and $|G_j|$ are relatively prime for $i \neq j$, then the least common multiple of $|G_1|, \ldots, |G_n|$ is $|G_1| \cdot \cdots \cdot |G_n|$. If we let g_i be a generator for G_i, then $o(g_i) = |G_i|$, so by part (i), $(g_1, \ldots, g_n)$ has order $|G_1| \cdot \cdots \cdot |G_n|$ in G. Thus $(g_1, \ldots, g_n)$ generates G, and G is cyclic. $\square$

Examples In $\mathbb{Z}_{12}$, $o(8) = 3$, and in $\mathbb{Z}_{18}$, $o(15) = 6$. Thus in $\mathbb{Z}_{12} \times \mathbb{Z}_{18}$, the order of the element $(8, 15)$ is the least common multiple of 3 and 6, namely 6.

The groups $\mathbb{Z}_{14} \times \mathbb{Z}_{15}$ and $\mathbb{Z}_8 \times \mathbb{Z}_9 \times \mathbb{Z}_5$ are cyclic; $\mathbb{Z}_{14} \times \mathbb{Z}_{16}$ and $\mathbb{Z}_8 \times \mathbb{Z}_9 \times \mathbb{Z}_6$ are not.

EXERCISES

6.1 Calculate the order of the element

a) $(4, 9)$ in $\mathbb{Z}_{18} \times \mathbb{Z}_{18}$.

b) $(7, 5)$ in $\mathbb{Z}_{12} \times \mathbb{Z}_8$.

c) $(8, 6, 4)$ in $\mathbb{Z}_{18} \times \mathbb{Z}_9 \times \mathbb{Z}_8$.

d) $(8, 6, 4)$ in $\mathbb{Z}_9 \times \mathbb{Z}_{17} \times \mathbb{Z}_{10}$.

6.2 Which of the following groups are cyclic?

a) $Z_{12} \times Z_9$

b) $Z_{10} \times Z_{85}$

c) $Z_4 \times Z_{25} \times Z_6$

d) $Z_{22} \times Z_{21} \times Z_{65}$

6.3 Is $Z \times Z$ cyclic? [Here Z means $(Z, +)$.]

6.4 Show that for finite groups $G_1, G_2, \ldots, G_n$,
$$|G_1 \times G_2 \times \cdots \times G_n| = |G_1| \cdot |G_2| \cdot \cdots \cdot |G_n|.$$

6.5 Let A be a subgroup of G, and let B be a subgroup of H. Show that $A \times B$ is a subgroup of $G \times H$.

6.6 Show that $G_1 \times G_2 \times \cdots \times G_n$ is abelian iff each G_i is abelian.

6.7 Construct a nonabelian group of order 16, and one of order 24.

6.8 Construct a group of order 81 with the property that every element except the identity has order 3.

6.9 Show that $Z(G_1 \times G_2 \times \cdots \times G_n) = Z(G_1) \times Z(G_2) \times \cdots \times Z(G_n)$.

6.10 Find all subgroups of $Z_2 \times Z_4$.

6.11 Find all subgroups of $Z_2 \times Z_2 \times Z_2$.

6.12 Let G and H be finite groups. Show that if $G \times H$ is cyclic, then (i) G and H are cyclic, and (ii) every subgroup of $G \times H$ is of the form $A \times B$ for some subgroups A and B of G and H, respectively.

6.13 Prove the converse of the result in Exercise 6.12; that is, show that for finite groups G and H, (i) and (ii) of Exercise 6.12 (taken together) imply that $G \times H$ is cyclic.

6.14 Use Theorem 6.1 to prove the **Chinese Remainder Theorem**: If $m_1, \ldots, m_n$ are positive integers such that m_i and m_j are relatively prime for $i \neq j$, and $k_1, \ldots, k_n$ are any integers, then there is an integer x such that $x \equiv k_i \pmod{m_i}$ for $1 \leq i \leq n$. (*Hint:* Consider the generator $(1, 1, \ldots, 1)$ for $Z_{m_1} \times \cdots \times Z_{m_n}$.)

FUNCTIONS

In this section we will present some elementary results about functions. These results will be useful in Section 8, when we investigate what are called *symmetric groups*, and later on, when we discuss homomorphisms and isomorphisms.

DEFINITION If S and T are sets, then a *function f from S to T* assigns to each $s \in S$ a unique element $f(s) \in T$.

As a definition this is somewhat strange, in that it tells you what a function *does* rather than what it *is*. Sometimes this difficulty is avoided by saying that a function is a "rule" that assigns elements of T to the elements of S, but this isn't any better because "rule" isn't defined. Besides, for some functions the "rule" is obscure at best, and it may be so hard to state that most people wouldn't call it a rule at all.

The above definition is fine as a working definition, and is how we usually think of functions. A more precise definition is as follows.

DEFINITION (precise) A **function** from S to T is a set of ordered pairs (s, t), where each $s \in S$ and each $t \in T$, such that each $s \in S$ occurs as the first element of one and only one pair (s, t).

Obviously this formulation captures the intent of our working definition; if $s \in S$, then there is only one pair (s, t) with s as its first element, and the function assigns the second element, t, of that pair to s. We write $f(s) = t$, and sometimes we say that f sends s to t, or maps s onto t. (The word "mapping" is sometimes used for "function.")

We write $f: S \to T$ to indicate that f is a function from S to T.

Examples

1. $S = T = \mathbb{R}$; $f: S \to T$ is given by $f(x) = x^2$.

2. $S = \{1,2\}$, $T = \{3,4,5\}$; $f: S \to T$ is given by $f(1) = 3$, $f(2) = 5$. In the precise formulation, f is $\{(1,3),(2,5)\}$.

3. $S = \mathbb{R}$, $T = [-1,1]$, the closed interval $-1 \leqslant x \leqslant 1$; $f: S \to T$ is given by $f(x) = \sin x$.

4. $S = GL(2,\mathbb{R})$, $T = \mathbb{R}$; $f: S \to T$ is given by

$$f\left(\begin{pmatrix} a & b \\ c & d \end{pmatrix}\right) = \text{determinant of } \begin{pmatrix} a & b \\ c & d \end{pmatrix} = ad - bc.$$

5. $S = $ set of continuous functions from $\mathbb{R}$ to $\mathbb{R}$, $T = \mathbb{R}$; $F: S \to T$ is given by $F(g) = \int_0^1 g(x)dx$.

6. $S = \mathbb{Z}_2 \times \mathbb{Z}_4$, $T = \mathbb{Z}_4$; $f: S \to T$ is given by $f((x,y)) = y$.

7. Let G be a group, and let $a \in G$. Define $f: G \to G$ by $f(x) = ax$. In the precise formulation, f is $\{(x,ax) | x \in G\}$.

8. If $S = \{1,2\}$, $T = \{3,4,5\}$, then $f = \{(1,3)\}$ is a function from $\{1\}$ to T, but it is not a function from S to T because $f(2)$ isn't defined. Also, $\{(1,3),(2,5),(1,4)\}$ isn't a function, because 1 appears as the first element of more than one pair.

Certain kinds of functions are particularly relevant to group theory.

DEFINITIONS Let $f: S \to T$ be a function. f is **onto** if for each $t \in T$ there is at least one $s \in S$ such that $f(s) = t$. f is **one-to-one** if whenever s_1 and s_2 are two different elements in S, we have $f(s_1) \neq f(s_2)$.

Thus f is onto iff every $t \in T$ comes from at least one $s \in S$; saying it yet another way, everything in T gets hit by f. You can think of f as a cannon firing shells (elements of S) at T (the target):

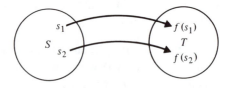

f is onto just if it doesn't miss anything in T.

On the other hand, f is one-to-one just if nothing in T gets hit twice. That is, anything in T is hit either just once or not at all.

Observe that if f is *both* one-to-one and onto, then every element of T is hit once and only once, so that f establishes a one-to-one correspondence between the elements of S and those of T: $t \in T$ is paired with the unique $s \in S$ such that $f(s) = t$.

There is some standard terminology that is good to know in connection with functions. If f: $S \to T$ is a function, then S is called the **domain** of f, and the set of elements of T which are hit by f (more precisely $\{t \in T|$ for some $s \in S, f(s) = t\})$ is called the **image** (or **range**) of f. Thus f is onto iff the image of f is T.

Examples Consider again Examples (1)–(7) on p. 60. The functions in Examples (3), (5), (6), and (7) are onto, and those in Examples (2) and (7) are one-to-one.

The function $f(x) = x^2$ in (1) is not onto since, for instance, $-1 \notin$ image of f. It is not one-to-one since $f(-1) = f(1)$.

The function in (4) is not onto since $0 \notin$ image of f, and it is not one-to-one since

$$f\left(\begin{pmatrix} 1 & 0 \\ 0 & 1 \end{pmatrix}\right) = f\left(\begin{pmatrix} 0 & 1 \\ -1 & 0 \end{pmatrix}\right).$$

The function F in (5) is onto since for any $r \in T$ we can find some $g \in S$ such that $F(g) = r$. For instance, the constant function g such that $g(x) = r$ for all x will do. F is not one-to-one since if g_1 and g_2 are defined by

$$g_1(x) = 0 \quad \text{for all } x, \quad g_2(x) = x - \tfrac{1}{2},$$

then g_1 and g_2 are different elements of S, but

$$F(g_1) = \int_0^1 0 \, dx = 0 = \int_0^1 \left(x - \frac{1}{2}\right) dx = F(g_2).$$

Finally, consider the function $f: G \to G$ in (7), given by $f(x) = ax$. f is onto since for every $y \in G$ there is some x such that $f(x) = y$. In fact, $x = a^{-1}y$ works, because then

$$f(x) = f(a^{-1}y) = a(a^{-1}y) = y.$$

This function is one-to-one by the left cancellation law: if $f(x_1) = f(x_2)$, that is, if $ax_1 = ax_2$, then $x_1 = x_2$. (Here we have used the definition of "one-to-one" in the following form: If $f: S \to T$, then f is one-to-one iff whenever $s_1, s_2 \in S$ and $f(s_1) = f(s_2)$, then $s_1 = s_2$.)

The term "injective" is sometimes used in place of "one-to-one." This terminology expresses the fact that the domain is carried intact into the image, without any collapsing taking place. People who use "injective" for "one-to-one" often use "surjective" for "onto"; this word indicates that the function throws the domain onto the set T.

Assume now that $f: S \to T$ is one-to-one and onto. As we have seen, f accomplishes a one-to-one pairing off of the elements of S with those of T. Therefore, f provides us with a function $f^{-1}: T \to S$, which maps any $t \in T$ onto the $s \in S$ with which t is paired by f. f^{-1} is called the **inverse function** of f.

Let us examine f^{-1} in a little more detail. Take any $t \in T$. Since f is onto, there exists at least one $s \in S$ such that $f(s) = t$. Since f is one-to-one, there is *only one* such s. Thus we can unambiguously define $f^{-1}(t)$ by setting $f^{-1}(t) = s$.

Observe that if f were not one-to-one, then there might be two different elements s_1 and s_2 of S such that both $f(s_1) = t$ and $f(s_2) = t$. We would then face a quandary in trying to define f^{-1}; the idea of the inverse function is that it is supposed to undo everything f did, and if $f(s_1) = f(s_2) = t$, then we cannot define $f^{-1}(t)$ so that f^{-1} undoes what f did to both s_1 and s_2. For example, if we were to define $f^{-1}(t) = s_1$, then f would send s_2 to t and f^{-1} would send t to s_1, rather than back to s_2. Similarly, if we were to define $f^{-1}(t) = s_2$, then f^{-1} would not undo what f did to s_1.

If f is one-to-one but not onto, then f has an inverse function f^{-1} with domain equal to the image of f. In order that f have an inverse function with domain T, it is thus necessary and sufficient that f be one-to-one and onto.

Observe that if $f: S \to T$ is one-to-one and onto and we view f as a set of ordered pairs (s, t), then $f^{-1}: T \to S$ is the function obtained by switching the entries in each pair in f: (s, t) is replaced by (t, s). It is not difficult to see that f^{-1} is itself both one-to-one and onto, so that it has an inverse $(f^{-1})^{-1}: S \to T$. In fact $(f^{-1})^{-1}$ is f, because we obtain it by switching all the pairs (t, s) back to (s, t).

Examples

1. Let $S = T = \mathbb{Z}$ and let $f: S \to T$ be given by $f(n) = n + 1$. Then f is one-to-one and onto, so it has an inverse function f^{-1}. In fact $f^{-1}(n) = n - 1$ for all $n \in \mathbb{Z}$.

2. Let $S = \mathbb{R}$ and $T = [-1, 1]$, and let $f: S \to T$ be given by $f(x) = \sin x$. Then f is onto but not one-to-one. (For example, $f(0) = f(\pi) = 0$.) We can restrict the domain of f so that f becomes one-to-one and onto, however. For example, if we restrict the domain to $[-\pi/2, \pi/2]$, then f has an inverse function which is called $\sin^{-1}$ or arcsin. This function is probably familiar to you from calculus.

3. Let $S = \mathbb{R}$ and $T = \mathbb{R}^+$, and let $f: S \to T$ be given by $f(x) = e^x$. Then f is one-to-one and onto, and its inverse f^{-1} is given by $f^{-1}(x) = \ln x$ for all $x \in \mathbb{R}^+$

The use of the term "inverse" is no accident vis-à-vis group theory. If X is a nonempty set, then we are going to define a binary operation on the set of all one-to-one onto functions $X \to X$, which turns this set into a group in such a way that the inverse of f in the above sense is the inverse of f in the group.

It will be useful to introduce this operation in a more general context, so let $f: S \to T$ and $g: T \to U$ be functions. We define the **composite function** $g \circ f: S \to U$ by setting

$$(g \circ f)(s) = g(f(s))$$

for all $s \in S$. Observe that $f(s) \in T$, so $g(f(s))$ makes sense and is an element of U.

Examples

1. Let $S = T = U = \mathbb{R}$, and let $f(x) = x^5 + 1$ and $g(x) = x^3$ for all $x \in \mathbb{R}$. Then

$$(g \circ f)(x) = g(x^5 + 1) = (x^5 + 1)^3,$$

for all $x \in \mathbb{R}$. What is $f \circ g$?

$$(f \circ g)(x) = f(x^3) = (x^3)^5 + 1 = x^{15} + 1.$$

Thus we see that $g \circ f$ and $f \circ g$ are not necessarily the same function.

In general we say that two functions f_1 and f_2 are *equal* and write $f_1 = f_2$ if f_1 and f_2 have the same domain and $f_1(x) = f_2(x)$ for all x in that domain. Thus in the present example we write $g \circ f \neq f \circ g$. Observe that in terms of the precise definition of a function, two functions are equal iff they are the same set of ordered pairs.

2. Let $f: S \to T$ be one-to-one and onto. Let i_S be the identity function on S, that is, $i_S(s) = s$ for all $s \in S$. Similarly, let i_T be the identity mapping on T. Then

$$f^{-1} \circ f: S \to S \quad \text{and} \quad f \circ f^{-1}: T \to T.$$

We have $f^{-1} \circ f = i_S$ and $f \circ f^{-1} = i_T$.

3. Let $S = T = \{(a,b) | a, b \in \mathbb{R}\}$. Let $f((a,b)) = (b,a)$, and let $g((a,b)) = (a, b+1)$. Then

$$(g \circ f)((a,b)) = (b, a+1)$$

and

$$(f \circ g)((a,b)) = (b+1, a).$$

Here again, $g \circ f \neq f \circ g$.

Now let X be a nonempty set, and let S_X denote the set of all one-to-one onto mappings $f: X \to X$.

THEOREM 7.1 $(S_X, \circ)$ is a group.

PROOF. If f and g are both in S_X, then certainly $f \circ g$ is a function from X to X. To check that $\circ$ is a binary operation on S_X, we must verify that if f, g are both one-to-one and onto, then $f \circ g$ is one-to-one and onto. This is left as an exercise.

Associativity of $\circ$ requires that if $f, g, h \in S_X$ then $(f \circ g) \circ h = f \circ (g \circ h)$. In other words, we must show that

$$[(f \circ g) \circ h](x) = [f \circ (g \circ h)](x)$$

for every $x \in X$. But

$$[(f \circ g) \circ h](x) = (f \circ g)(h(x)) = f(g(h(x))),$$

and

$$[f \circ (g \circ h)](x) = f((g \circ h)(x)) = f(g(h(x))).$$

The identity element of $(S_X, \circ)$ is i_X. That is, $f \circ i_X = i_X \circ f = f$, for every $f \in S_X$.

Finally, if $f \in S_X$ and f^{-1} denotes its inverse function, then $f^{-1} \in S_X$ and

$$f \circ f^{-1} = f^{-1} \circ f = i_X,$$

so f^{-1} is the group-theoretic inverse of f. $\square$

EXERCISES

7.1 In each example below, f is given either as a rule or as a set of ordered pairs. In each case, determine whether or not f is a function from S to T. For those cases in which it is, determine whether it is one-to-one, and whether it is onto.

a) $S = \{1, 2, 3, 4, 5\}$, $T = \{6, 7, 8, 9, 10\}$, $f = \{(1, 8), (3, 9), (4, 10), (2, 6), (5, 9)\}$

b) S and T are as in (a), $f = \{(1, 8), (3, 10), (2, 6), (4, 9)\}$

c) S and T are as in (a), $f = \{(1, 7), (2, 6), (4, 5), (1, 9), (5, 10)\}$

d) $S = T = \mathbb{R}$, $f(x) = x^2 - x$

e) $S = T = \mathbb{R}$, $f(x) = x^3$

f) $S = T = \mathbb{R}$, $f(x) = \sqrt{x}$

g) $S = T = \mathbb{R}$, $f(x) = 1/x$

h) $S = T = \mathbb{Z}^+$, $f(x) = x + 1$

i) $S = T = \mathbb{Z}^+$, $f(x) = \begin{cases} 1 & \text{if } x = 1 \\ x - 1 & \text{if } x > 1 \end{cases}$

j) $S = T = \mathbb{R}^+$, $f(x) = \dfrac{x}{x^2 + 1}$

7.2 Let $f : \mathbb{R} \to \mathbb{R}$.

a) Give a condition on the graph of $y = f(x)$, in terms of its intersections with horizontal lines, that is equivalent to f being one-to-one.

b) If $g : \mathbb{R} \to \mathbb{R}$ and f and g are both one-to-one, must $f + g$ be one-to-one?

7.3 Let $f: \mathbb{R} \to \mathbb{R}$ be given by $f(x) = ax + b$, where a and b are fixed constants.

    a) Show that if $a \neq 0$ then f is one-to-one and onto, so that f^{-1} exists.

    b) Assuming that $a \neq 0$, find an explicit formula for the inverse function f^{-1}.

7.4 Let $S = T =$ the set of polynomials with real coefficients, and define a function from S to T by mapping each polynomial to its derivative. Is this function one-to-one? Is it onto?

7.5 Let X be a set, and let $A \subseteq X$. Define a function $f: P(X) \to P(X)$ by $f(B) = A \cap B$, for $B \in P(X)$. Under what conditions is f one-to-one and onto?

7.6 Let X be a set, and let $A \subseteq X$. Define $f: P(X) \to P(X)$ by $f(B) = A \triangle B$. Is f one-to-one? Is f onto?

7.7 Let G be a group and let $a \in G$. Define a function $f: G \to G$ by $f(x) = axa^{-1}$ for all $x \in G$. Is f one-to-one? Is f onto?

7.8 Let G be a group, and let $f(x) = x^{-1}$ for all $x \in G$. Is f a function from G to G? If so, is it one-to-one? Is it onto?

7.9 Show that $\circ$ is a binary operation in Theorem 7.1.

7.10 Let $f: S \to T$.

    a) Show that f is one-to-one if and only if there exists a function $g: T \to S$ such that $g \circ f = i_S$.

    b) Show that f is onto if and only if there exists a function $g: T \to S$ such that $f \circ g = i_T$.

    c) Show that f is one-to-one and onto if and only if there exists a function $g: T \to S$ such that $g \circ f = i_S$ and $f \circ g = i_T$.

7.11 Let $f: S \to T$ and $g: T \to U$.

    a) If $g \circ f$ is one-to-one, must both f and g be one-to-one?

    b) If $g \circ f$ is onto, must both f and g be onto?

7.12 Let $f: S \to T$. For any subset A of S, define

$$f(A) = \{ f(s) \mid s \in A \}.$$

    a) Show that if A, B are subsets of S, then $f(A \cup B) = f(A) \cup f(B)$.

    b) Show that $f(A \cap B) \subseteq f(A) \cap f(B)$. Construct an example where the inclusion is proper, i.e., $f(A \cap B) \subsetneq f(A) \cap f(B)$.

SYMMETRIC GROUPS

If X is a nonempty set, then a one-to-one onto mapping $X \to X$ is called a **permutation** of X. We have seen that the set of all such permutations forms a group $(S_X, \circ)$ under composition of functions.

DEFINITION $(S_X, \circ)$ is called the **symmetric group on X**.

Symmetric groups were used in mathematics before the abstract concept of "group" had been formulated. In particular, they were used to obtain deep and incisive results about the solutions of polynomial equations, and the success of these efforts gave an impetus to the development of the abstract theory. After the abstract notion was established, the English mathematician Arthur Cayley (1821–1895) again demonstrated the importance of symmetric groups by showing that *every* group can be thought of as a subgroup of some symmetric group. We will state Cayley's result precisely, and prove it, in Section 12.

For now, we will try to get familiar with symmetric groups by investigating symmetric groups on finite sets. If X is finite and has, say, n elements, then we can represent X by $\{1, 2, \ldots, n\}$, and we accordingly denote $(S_X, \circ)$ by S_n in this case. S_n is called the **symmetric group of degree n**.

Let $f \in S_n$. Then f shuffles the elements $1, 2, 3, \ldots, n$, and we can represent f explicitly by writing

$$\begin{pmatrix} 1 & 2 & 3 & \cdots & n \\ f(1) & f(2) & f(3) & \cdots & f(n) \end{pmatrix},$$

where $f(k)$ is placed under k for each k between 1 and n.

It is easy to calculate products in S_n. For example, consider

$$\begin{pmatrix} 1 & 2 & 3 & 4 \\ 2 & 4 & 1 & 3 \end{pmatrix} \circ \begin{pmatrix} 1 & 2 & 3 & 4 \\ 3 & 2 & 4 & 1 \end{pmatrix}$$

in S_4. To see what goes under 1 in the product we just recall that the product is the composition of the two given permutations, the one on the right being performed first:

$$\begin{pmatrix} 1 & 2 & 3 & 4 \\ 2 & 4 & 1 & 3 \end{pmatrix} \circ \begin{pmatrix} 1 & 2 & 3 & 4 \\ 3 & 2 & 4 & 1 \end{pmatrix}(1) = \begin{pmatrix} 1 & 2 & 3 & 4 \\ 2 & 4 & 1 & 3 \end{pmatrix}(3) = 1.$$

Similarly,

$$\begin{pmatrix} 1 & 2 & 3 & 4 \\ 2 & 4 & 1 & 3 \end{pmatrix} \circ \begin{pmatrix} 1 & 2 & 3 & 4 \\ 3 & 2 & 4 & 1 \end{pmatrix}(2) = 4,$$

because 2 goes to 2, which then goes to 4. The product is

$$\begin{pmatrix} 1 & 2 & 3 & 4 \\ 2 & 4 & 1 & 3 \end{pmatrix} \circ \begin{pmatrix} 1 & 2 & 3 & 4 \\ 3 & 2 & 4 & 1 \end{pmatrix} = \begin{pmatrix} 1 & 2 & 3 & 4 \\ 1 & 4 & 3 & 2 \end{pmatrix}.$$

Observe that the notation for the product

$$\begin{pmatrix} 1 & 2 & 3 & 4 \\ 1 & 4 & 3 & 2 \end{pmatrix}$$

is rather uneconomical. Nothing happens to 1 or 3 (they are left fixed), and the whole permutation does nothing more than interchange 2 and 4. To achieve a more efficient notation for permutations—and to introduce an important subgroup of S_n—we consider special permutations called **cycles**.

Let $x_1, x_2, \ldots, x_r$, $1 \le r \le n$, be r distinct elements of $\{1, 2, \ldots, n\}$. The **r-cycle** $(x_1, x_2, \ldots, x_r)$ is the element of S_n that maps $x_1 \to x_2$, $x_2 \to x_3, \ldots, x_{r-1} \to x_r$, $x_r \to x_1$, and leaves all elements of $\{1, 2, 3, \ldots, n\}$ other than $x_1, x_2, \ldots, x_r$ fixed:

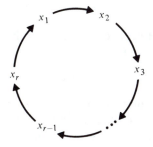

This cycle could just as well be written

$$(x_r, x_1, x_2, \ldots, x_{r-1}) \quad \text{or} \quad (x_{r-1}, x_r, x_1, \ldots, x_{r-2}), \text{ etc.}$$

For example,

$$\begin{pmatrix} 1 & 2 & 3 & 4 \\ 1 & 4 & 3 & 2 \end{pmatrix}$$

can be written more simply as $(2, 4)$, (or $(4, 2)$, which is the same thing). The identity permutation

$$\begin{pmatrix} 1 & 2 & 3 & 4 & 5 \\ 1 & 2 & 3 & 4 & 5 \end{pmatrix}$$

in S_5 can be written as (1), or as (2), or (3), or (4), or (5). It can also be written as $(2,5) \circ (2,5)$, since the right-hand factor just switches 2 and 5, and the left-hand factor switches them back.

Two cycles $(x_1, x_2, \ldots, x_r)$ and $(y_1, y_2, \ldots, y_s)$ in S_n are called **disjoint** if no element of $\{1, 2, \ldots, n\}$ is moved by both cycles. If $r \geq 2$ and $s \geq 2$ this can be expressed by saying that

$$\{x_1, x_2, \ldots, x_r\} \cap \{y_1, y_2, \ldots, y_s\} = \varnothing.$$

It is not difficult to see that every permutation can be written as the product of a finite number of cycles, any two of which are disjoint.

THEOREM 8.1 Let $f \in S_n$. Then there exist disjoint cycles $f_1, f_2, \ldots, f_m$ in S_n such that $f = f_1 \circ f_2 \circ \cdots \circ f_m$.

PROOF. Choose some $x_1 \in \{1, 2, \ldots, n\}$. Let $x_2 = f(x_1), x_3 = f(x_2)$, and so on. Since $\{1, 2, 3, \ldots, n\}$ is a finite set, there must be a first element in the sequence $x_1, x_2, x_3, \ldots$ which is the same as a previous element. Say this element is x_k and $x_k = x_j, j < k$. Then j must be 1, since if $j > 1$, $x_k = x_j$ implies $x_{k-1} = x_{j-1}$ (since f is one-to-one), contradicting the minimality of k. Thus the first $k-1$ elements of the sequence $x_1, x_2, x_3, \ldots$ are distinct and $x_k = x_1$. Thus f includes the cycle $f_1 = (x_1, x_2, \ldots, x_{k-1})$, and

$$f = f_1 \circ h_1,$$

where h_1 permutes the elements of $\{1, 2, \ldots, n\}$ other than $x_1, \ldots, x_{k-1}$. Repeating our argument on h_1, we write

$$h_1 = f_2 \circ h_2,$$

where f_2 is a cycle disjoint from f_1, and h_2 permutes the elements of $\{1, 2, \ldots, n\}$ not contained in either f_1 or f_2. If we continue this process long enough, we must come to a point where h_m has nothing left to permute, that is, h_m is the identity permutation. Then

$$f = f_1 \circ f_2 \circ f_3 \circ \cdots \circ f_m \circ h_m = f_1 \circ f_2 \circ \cdots \circ f_m. \quad \square$$

Example Consider the element

$$f = \begin{pmatrix} 1 & 2 & 3 & 4 & 5 & 6 & 7 & 8 \\ 3 & 5 & 7 & 4 & 2 & 8 & 1 & 6 \end{pmatrix}$$

in S_8. We have

$$f = (1,3,7) \circ \begin{pmatrix} 1 & 2 & 3 & 4 & 5 & 6 & 7 & 8 \\ 1 & 5 & 3 & 4 & 2 & 8 & 7 & 6 \end{pmatrix}$$

$$= (1,3,7) \circ (2,5) \circ \begin{pmatrix} 1 & 2 & 3 & 4 & 5 & 6 & 7 & 8 \\ 1 & 2 & 3 & 4 & 5 & 8 & 7 & 6 \end{pmatrix}$$

$$= (1,3,7) \circ (2,5) \circ (6,8).$$

We might also have written this as $(1,3,7) \circ (2,5) \circ (4) \circ (6,8)$, but (4) is just the identity element, and it is more economical to omit it.

The factorization of permutations into disjoint cycles is very much like the factorization of integers into primes in the Fundamental Theorem of Arithmetic. It is easy to see that disjoint cycles commute (Exercise 8.8), so that once we have

$$f = f_1 \circ f_2 \circ f_3 \circ \cdots \circ f_m,$$

we can also write

$$f = f_2 \circ f_1 \circ f_3 \circ \cdots \circ f_m,$$

and so on. However, if we omit all factors which are the identity, then the factorization is unique, except for this ability to rearrange the factors.

A 2-cycle, i.e., a cycle that just interchanges two elements, is called a **transposition**. We continue our decomposition of permutations by proving

THEOREM 8.2 If $n \geqslant 2$, then any cycle in S_n can be written as a product of transpositions.

PROOF. A 1-cycle is the identity, hence can be written as $(1,2) \circ (1,2)$. For an r-cycle with $r \geqslant 2$, we have

$$(x_1, x_2, \ldots, x_r) = (x_1, x_r) \circ (x_1, x_{r-1}) \circ (x_1, x_{r-2}) \circ \cdots \circ (x_1, x_3) \circ (x_1, x_2). \quad \square$$

Example Referring to the previous example, we have $(1,3,7) = (1,7) \circ (1,3)$. We also have $(1,3,7) = (4,7) \circ (1,7) \circ (1,4) \circ (1,3)$, and in general there are many ways in which a cycle can be written as a product of transpositions. We have now lost the uniqueness of factorization.

Combining the last two theorems proves

THEOREM 8.3 If $n \geqslant 2$, then any element of S_n can be written as a product of transpositions.

Example Referring again to the example of

$$f = \begin{pmatrix} 1 & 2 & 3 & 4 & 5 & 6 & 7 & 8 \\ 3 & 5 & 7 & 4 & 2 & 8 & 1 & 6 \end{pmatrix} \in S_8,$$

we have

$$f = (1,7) \circ (1,3) \circ (2,5) \circ (6,8),$$

and also

$$f = (4,7) \circ (1,7) \circ (1,4) \circ (1,3) \circ (2,5) \circ (6,8).$$

DEFINITION A permutation is **even** if it can be written as the product of an even number of transpositions. It is **odd** if it can be written as the product of an odd number of transpositions.

We would not want it to be possible for a permutation to be both even and odd, and it turns out that it *isn't* possible. Although a given permutation may have many representations as products of transpositions, it will always be the case that either these products all have an even number of factors or they all have an odd number of factors. The neatest, most natural proof of this fact that we have seen is a relatively new proof, published in 1971 by William Miller ("Even and Odd Permutations," *Mathematics Association of Two-Year Colleges Journal*, vol. 5, p. 32). Here it is:

THEOREM 8.4 No permutation is both even and odd.

PROOF. Suppose that f is both even and odd. Then we have

$$f = t_1 t_2 \cdots t_k = s_1 s_2 \cdots s_\ell,$$

where k is even, ℓ is odd, and the t's and s's are transpositions. Thus

$$t_1 t_2 \cdots t_k \cdot s_\ell^{-1} \cdot s_{\ell-1}^{-1} \cdots s_2^{-1} s_1^{-1} = \text{identity},$$

that is, $t_1 t_2 \cdots t_k s_\ell s_{\ell-1} \cdots s_2 s_1 = \text{identity},$

so since $k + \ell$ is odd, the identity permutation is written as the product of an odd number of transpositions. We are going to show that this is impossible, hence f could not have been both even and odd.

Consider for a moment an arbitrary product $P = t_1 t_2 \cdots t_m$ of transpositions. We assert that either

(I) P is not the identity permutation,   or

(II) P equals a product of $m - 2$ transpositions.

To see this, let a be some element of $\{1, 2, \ldots, n\}$ that occurs in some t, and let t_j be the first t *from the right* in which a appears. Say t_j is (a, b).

Now if t_j is t_1, we have (I), since then P does not fix a. If t_j is not t_1, then consider t_{j-1}. It must be either:

i) (a,b), in which case $t_{j-1}t_j = $ identity;
ii) (a,c), for some $c \neq b$, in which case $t_{j-1}t_j = (a,c)(a,b) = (a,b)(b,c)$;
iii) (b,c), for some $c \neq a$, in which case $t_{j-1}t_j = (b,c)(a,b) = (a,c)(b,c)$;
iv) (c,d) for some $c \neq a,b$ and $d \neq a,b$, in which case $t_{j-1}t_j = (c,d)(a,b) = (a,b)(c,d)$.

If Case (i) occurs, then we can delete $t_{j-1}t_j$ from P, and we have (II). If any of the other cases occurs, then by using the indicated equalities we can find an expression equal to P in which the rightmost occurrence of a is one factor further to the left than it was when we started.

If we now keep repeating our argument on a, then either Case (i) eventually occurs, and we have (II), or else we eventually replace P by an equivalent expression in which the rightmost occurrence of a is in the leftmost factor, and we have (I) as above.

Thus we have proved our assertion about P. It now follows that if m is *odd*, then $P = t_1 t_2 \cdots t_m$ cannot be the identity permutation. For if it is, then (II) must hold for P, so we can write the identity as a product of $m-2$ transpositions. Then (II) holds for this shorter factorization, so we can find another factorization with $m-4$ factors. Continuing in this way, we eventually reduce the number of factors to 1 (since m was odd to start with), and then we are done because it is clear that a single transposition is not the identity. $\square$

You might enjoy writing down a few products of transpositions for yourself, and seeing how either (I) or (II) comes true for them.

Now let $n \geq 2$, and let A_n denote the subset of S_n consisting of all the even permutations. Recall that if n is a positive integer, then $n!$ (read "n factorial") denotes the product $n(n-1)(n-2)\cdots(3)(2)(1)$.

THEOREM 8.5 Let $n \geq 2$. Then A_n is a subgroup of S_n. $|S_n| = n!$, and $|A_n| = n!/2$.

PROOF. To see that $|S_n| = n!$, we consider what it takes to determine an element $f \in S_n$. We must choose $f(1)$ from among $\{1, 2, \ldots, n\}$; for any choice we make, there are $n-1$ ways of choosing $f(2)$, because we must choose $f(2)$ from the set $\{1, 2, \ldots, n\} - \{f(1)\}$. Thus there are $n(n-1)$ ways of determining what $f(1)$ and $f(2)$ are going to be. For any one of these ways, there are $n-2$ ways of choosing $f(3)$, so there are $n(n-1)(n-2)$ ways of choosing the first three values of f. Continuing in this way, we see that there are $n!$ ways of specifying an element of S_n.

Now to the claims about A_n. A_n is a nonempty subset of S_n and therefore a nonempty finite subset of S_n. (Subsets of finite sets are finite.) Since the product of two even permutations is even (why?), A_n is closed under the group operation in S_n. Therefore, A_n is a subgroup of S_n by Theorem 5.3.

To see that $|A_n| = n!/2$, we want to show that A_n contains exactly one half of the elements of S_n. Observe that if $f_1, f_2, \ldots, f_k$ are all the distinct elements of A_n, then it suffices to show that there are exactly k distinct *odd* permutations in S_n. If we let g be $(1,2)$, then

$$gf_1, gf_2, \ldots, gf_k$$

are all distinct and all odd. (Why?) Furthermore, these are *all* the odd permutations in S_n. For if h is odd, then gh is even, so gh is one of $f_1, f_2, \ldots, f_k$ and

$$g(gh) \in \{ gf_1, gf_2, \ldots, gf_k \}.$$

But $g(gh)$ is h, since g^2 is the identity. Therefore, every odd permutation is one of $gf_1, gf_2, \ldots, gf_k$, and we have shown that there are exactly k odd permutations in S_n, as desired. $\square$

A_n is called the **alternating group of degree n**. Alternating groups (especially A_4) will be useful to us a little later on.

We conclude this section by working out some examples in detail.

First let's look at S_3. By Theorem 8.5, S_3 has $3! = 6$ elements. They are

$$\begin{pmatrix} 1 & 2 & 3 \\ 1 & 2 & 3 \end{pmatrix} = e,$$

$$\begin{pmatrix} 1 & 2 & 3 \\ 2 & 3 & 1 \end{pmatrix} = (1,2,3),$$

$$\begin{pmatrix} 1 & 2 & 3 \\ 3 & 1 & 2 \end{pmatrix} = (1,3,2),$$

$$\begin{pmatrix} 1 & 2 & 3 \\ 1 & 3 & 2 \end{pmatrix} = (2,3),$$

$$\begin{pmatrix} 1 & 2 & 3 \\ 2 & 1 & 3 \end{pmatrix} = (1,2),$$

$$\begin{pmatrix} 1 & 2 & 3 \\ 3 & 2 & 1 \end{pmatrix} = (1,3).$$

If we denote the second element in the list by f, then

$$f^2 = \begin{pmatrix} 1 & 2 & 3 \\ 2 & 3 & 1 \end{pmatrix}\begin{pmatrix} 1 & 2 & 3 \\ 2 & 3 & 1 \end{pmatrix} = \begin{pmatrix} 1 & 2 & 3 \\ 3 & 1 & 2 \end{pmatrix},$$

and

$$f^3 = \begin{pmatrix} 1 & 2 & 3 \\ 3 & 1 & 2 \end{pmatrix}\begin{pmatrix} 1 & 2 & 3 \\ 2 & 3 & 1 \end{pmatrix} = e.$$

Thus $o(f)=3$, and f generates the cyclic subgroup $\langle f \rangle = \{e, f, f^2\}$. If we let

$$g = \begin{pmatrix} 1 & 2 & 3 \\ 1 & 3 & 2 \end{pmatrix},$$

then

$$g^2 = \begin{pmatrix} 1 & 2 & 3 \\ 1 & 3 & 2 \end{pmatrix}\begin{pmatrix} 1 & 2 & 3 \\ 1 & 3 & 2 \end{pmatrix} = e,$$

so $o(g)=2$ and $\langle g \rangle = \{e,g\}$. The only subgroup of S_3 containing both f and g is S_3 itself, because

$$fg = \begin{pmatrix} 1 & 2 & 3 \\ 2 & 3 & 1 \end{pmatrix}\begin{pmatrix} 1 & 2 & 3 \\ 1 & 3 & 2 \end{pmatrix} = \begin{pmatrix} 1 & 2 & 3 \\ 2 & 1 & 3 \end{pmatrix}$$

and

$$f^2g = \begin{pmatrix} 1 & 2 & 3 \\ 3 & 1 & 2 \end{pmatrix}\begin{pmatrix} 1 & 2 & 3 \\ 1 & 3 & 2 \end{pmatrix} = \begin{pmatrix} 1 & 2 & 3 \\ 3 & 2 & 1 \end{pmatrix}.$$

Thus

$$S_3 = \{e, f, f^2, g, fg, f^2g\}.$$

Of course, gf is also an element of S_3, and in fact

$$gf = \begin{pmatrix} 1 & 2 & 3 \\ 1 & 3 & 2 \end{pmatrix}\begin{pmatrix} 1 & 2 & 3 \\ 2 & 3 & 1 \end{pmatrix} = \begin{pmatrix} 1 & 2 & 3 \\ 3 & 2 & 1 \end{pmatrix} = f^2g.$$

Likewise, $gf^2 = fg$.

The equations $gf = f^2g$ and $gf^2 = fg$ tell us how to multiply the elements of S_3 without writing down the permutation notation. For instance,

$$(fg)(fg) = f(gf)g = f(f^2g)g = f^3g^2 = e,$$

and

$$(f^2g)(f^2g) = f^2(gf^2)g = f^2(fg)g = f^3g^2 = e,$$

so both fg and f^2g have order 2. Observe that in these calculations we simply used the equations $gf = f^2g$ and $gf^2 = fg$ to "move f's past g," and thus get all the f's together and all the g's together.

Let's figure out what the subgroups of S_3 are. Of course we have the cyclic subgroups $\langle e \rangle, \langle f \rangle, \langle f^2 \rangle, \langle g \rangle, \langle fg \rangle$, and $\langle f^2g \rangle$. Since $\langle f \rangle = \langle f^2 \rangle$ and g, fg, f^2g are each of order 2, there is one cyclic subgroup of order 3, and there are 3 cyclic subgroups of order 2. How about other subgroups?

As we have seen, any subgroup of S_3 that contains f and g must be all of S_3. Suppose that a subgroup contains f but not g. Then it cannot contain fg, because if it did it would contain $f^2(fg)=g$. Similarly, it cannot contain f^2g, so it is $\langle f \rangle$. Next, if a subgroup contains g but not f, then it must be $\langle g \rangle$, since if it contained f^2, it would contain $(f^2)^2=f$; and if it contained fg or f^2g, then it would contain $(fg)g=f$ or $(f^2g)g=f^2$.

Finally, consider the possibilities for a subgroup that contains neither f nor g. It cannot contain f^2, and therefore it cannot contain both fg and f^2g, because their product is

$$(fg)(f^2g) = f(gf^2)g = f(fg)g = f^2.$$

If it contains neither fg nor f^2g, then it is the trivial subgroup $\langle e \rangle$; otherwise it is $\langle fg \rangle$ or $\langle f^2g \rangle$.

Thus the only proper subgroups of S_3 are the cyclic subgroups given above. The subgroup lattice for S_3 looks like this:

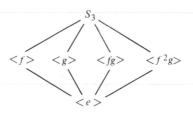

By the way, we know that one of the subgroups must be A_3, and since A_3 must have order 3, by Theorem 8.5, we have $A_3 = \langle f \rangle$. You can also check directly that $\langle f \rangle$ consists precisely of the even permutations in S_3.

One particularly interesting thing about S_3 is that it is nonabelian ($fg \neq gf$) and has fewer elements than any other nonabelian group we have seen. We will show in Section 10 that every group of order ≤ 5 is abelian, so S_3 is in fact the smallest nonabelian group.

We will often use S_3 as an example, so it is probably worthwhile for you to become familiar with it now. The structure of S_3 is easily remembered by thinking in terms of f and g. (You may find it easier to remember the equations $gf = f^2g$ and $gf^2 = fg$ in the form $gf^i = f^{-i}g$, $i = 1, 2$.)

As our final example, we will consider a subgroup of S_4 that arises from geometric considerations. Suppose we have a square P_4 drawn in a plane, and we have labeled the corners as "places":

Place 1 ┌─────┐ Place 2

Place 4 └─────┘ Place 3

Suppose we also have a cardboard square the same size as P_4, placed on the plane figure, with its vertices labeled A, B, C, and D:

Place 1 │ A   B │ Place 2

Place 4 │ D   C │ Place 3

By a **symmetry** of P_4, we mean any resituation of the cardboard square that can be accomplished by picking it up, rotating it and/or flipping it over any number of times, and then placing it back on the plane figure. There are exactly eight symmetries, because there are four positions we can achieve without using flips, and four more we can get by using one flip.

Any symmetry of P_4 gives us an element of S_4, because for each j, $1 < j < 4$, it tells us to move the vertex in place j to some place j'. We obtain our element of S_4 by mapping each j to the corresponding j'. For instance, a $90°$ clockwise rotation of the cardboard square gives us

$$\begin{pmatrix} 1 & 2 & 3 & 4 \\ 2 & 3 & 4 & 1 \end{pmatrix} = (1,2,3,4),$$

and a flip across the diagonal from upper left to lower right gives us

$$\begin{pmatrix} 1 & 2 & 3 & 4 \\ 1 & 4 & 3 & 2 \end{pmatrix} = (2,4).$$

Suppose we take symmetries F_1 and F_2, which give us permutations f_1 and f_2, respectively. The symmetry $F_2 \circ F_1$ tells us to take the vertex in place j, move it to place j' as determined by F_1, and from there to place j'' as determined by F_2. Thus $F_2 \circ F_1$ gives us the permutation $f_2 \circ f_1$.

Now let D_4 denote the subset of S_4 consisting of the eight permutations that come from the symmetries of P_4. Then D_4 is a finite subset of S_4, and it is closed under multiplication in S_4, by the observation in the preceding paragraph. Hence by Theorem 5.3, D_4 is a subgroup of S_4. It is called the **octic group**, or **group of symmetries of a square.**

If we let f be the $90°$ clockwise rotation $(1,2,3,4)$, and g the flip $(2,4)$, then $o(f) = 4$ and $o(g) = 2$. The four elements of $\langle f \rangle = \{e, f, f^2, f^3\}$ correspond to the four positions we can attain without using flips, and the four elements g, fg, f^2g, f^3g correspond to the positions that require a flip. Thus

$$D_4 = \{e, f, f^2, f^3, g, fg, f^2g, f^3g\}.$$

As for S_3, we can multiply these elements without resorting to permutation notation once we know how to move f's past g. Now $gfg = f^{-1}$, a counterclockwise rotation of $90°$. (Convince yourself!) Thus for any integer i, $(gfg)^i = f^{-i}$, and since $g^2 = e$, this becomes $gf^ig = f^{-i}$. Thus

$$gf^i = f^{-i}g$$

(just as for S_3!), and we are all set. For instance,

$$(f^2g)f^3 = f^2(gf^3) = f^2(f^{-3}g) = f^{-1}g = f^3g.$$

Likewise, for $i = 0, 1, 2, 3$, we have

$$(f^ig)^2 = (f^ig)(f^ig) = f^if^{-i}g^2 = e,$$

in accordance with the fact that the elements $f^i g$ correspond to flips about diagonals or vertical or horizontal axes, and thus have order 2 on geometric grounds.

An investigation of the subgroups of D_4, analogous to the one we carried out for S_3, yields the following lattice:

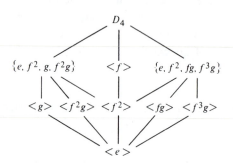

EXERCISES

8.1 Carry out the indicated multiplications in S_6.

a) $\begin{pmatrix} 1 & 2 & 3 & 4 & 5 & 6 \\ 3 & 6 & 1 & 4 & 2 & 5 \end{pmatrix} \circ \begin{pmatrix} 1 & 2 & 3 & 4 & 5 & 6 \\ 5 & 4 & 3 & 2 & 1 & 6 \end{pmatrix}$

b) $\begin{pmatrix} 1 & 2 & 3 & 4 & 5 & 6 \\ 3 & 2 & 5 & 4 & 1 & 6 \end{pmatrix} \circ \begin{pmatrix} 1 & 2 & 3 & 4 & 5 & 6 \\ 6 & 5 & 4 & 1 & 2 & 3 \end{pmatrix}$

c) $\begin{pmatrix} 1 & 2 & 3 & 4 & 5 & 6 \\ 2 & 4 & 6 & 1 & 3 & 5 \end{pmatrix} \circ \begin{pmatrix} 1 & 2 & 3 & 4 & 5 & 6 \\ 5 & 6 & 3 & 4 & 1 & 2 \end{pmatrix}$

8.2 Write each permutation as a product of disjoint cycles, and then as a product of transpositions. Determine whether each permutation is even or odd.

a) $\begin{pmatrix} 1 & 2 & 3 & 4 & 5 & 6 \\ 3 & 6 & 1 & 4 & 2 & 5 \end{pmatrix}$

b) $\begin{pmatrix} 1 & 2 & 3 & 4 & 5 & 6 \\ 2 & 4 & 6 & 1 & 3 & 5 \end{pmatrix}$

c) $\begin{pmatrix} 1 & 2 & 3 & 4 & 5 & 6 \\ 5 & 6 & 3 & 4 & 1 & 2 \end{pmatrix}$

d) $\begin{pmatrix} 1 & 2 & 3 & 4 & 5 & 6 \\ 6 & 5 & 4 & 1 & 2 & 3 \end{pmatrix}$

8.3 a) Let $f = (x_1, x_2, x_3, \ldots, x_r)$ be a cycle. For which r's is f an even permutation?

b) How do you find out whether a given permutation is even or odd without factoring it into transpositions?

c) Determine whether

$$\begin{pmatrix} 1 & 2 & 3 & 4 & 5 & 6 & 7 & 8 & 9 & 10 & 11 & 12 \\ 6 & 7 & 5 & 9 & 8 & 4 & 11 & 3 & 1 & 12 & 2 & 10 \end{pmatrix} \in S_{12}$$

is even or odd.

8.4 Let $f=(x_1,x_2,\ldots,x_r)\in S_n$. Show that $o(f)=r$.

8.5 Write down all the elements of S_4, and indicate which ones are in A_4. Check your results against Theorem 8.5.

8.6 Show that there is no element $f\in S_5$ such that $f(1,2,3)f^{-1}=(1,2)(3,4,5)$.

8.7 Prove that S_n is nonabelian if $n\geqslant 3$.

8.8 Show that if f and g are disjoint cycles, then $fg=gf$.

8.9 Verify the uniqueness of the decomposition of permutations into disjoint cycles.

8.10 (See Exercises 8.4 and 8.8.)

a) Suppose $f,g\in S_n$ are disjoint cycles, $o(f)=r$, and $o(g)=s$. Show that $o(fg)=$ the least common multiple of r and s.

b) Suppose that a permutation f is the product of disjoint cycles $f_1,f_2,\ldots,f_n$. Show that $o(f)$ is the least common multiple of $o(f_1),o(f_2),\ldots,o(f_n)$.

c) Find the order of

$$\begin{pmatrix} 1 & 2 & 3 & 4 & 5 & 6 & 7 & 8 & 9 & 10 & 11 & 12 \\ 6 & 7 & 5 & 9 & 8 & 4 & 11 & 3 & 1 & 12 & 2 & 10 \end{pmatrix}$$

in S_{12}.

8.11 Find:

a) $Z(S_3)$;

b) $Z(D_4)$.

8.12 Let $n>3$, and let P_n be a regular polygon with n sides. By using

$$f=(1,2,\ldots,n) \quad \text{and} \quad g=\begin{pmatrix} 1 & 2 & 3 & 4 & \ldots & n-2 & n-1 & n \\ 1 & n & n-1 & n-2 & \ldots & 4 & 3 & 2 \end{pmatrix},$$

we can repeat our derivation of D_4 from P_4 and obtain a subgroup D_n of S_n. We get

$$D_n=\{e,f,f^2,\ldots,f^{n-1},g,fg,f^2g,\ldots,f^{n-1}g\},$$

where $o(f)=n$, $o(g)=2$, and $gf^i=f^{-i}g$ for all i. D_n is called **the dihedral group of order $2n$**.

a) Show that $D_3=S_3$, but $D_n\neq S_n$ for $n>4$.

b) Calculate $(f^2g)(f)(f^3g)$ in D_5.

c) Find $Z(D_5)$.

8.13 Let X be a set, and let $Y\subseteq X$. Show that the subset of S_X consisting of all f such that $f(y)=y$ for all $y\in Y$ forms a subgroup.

8.14 For which n, $n>2$, do the cycles in S_n form a subgroup? Do the odd permutations form a subgroup?

8.15 We have seen that in an abelian group, the set of elements x such that $x^n=e$ (where n is a fixed positive integer) forms a subgroup. Give an example of a group such that for some n, the elements whose nth power is e do *not* form a subgroup.

8.16 Let f, g be distinct transpositions.

a) Show that if f and g are disjoint, then fg can be expressed as the product of two 3-cycles.

b) Show that if f and g are not disjoint, then fg can be expressed as a 3-cycle.

8.17 Prove that, for $n > 3$, every even permutation in S_n can be expressed as a product of 3-cycles. (Thus A_n is "generated" by the 3-cycles. See Exercise 8.16.)

8.18 Find elements $x, y \in S_{\mathbf{Z}}$ such that x and y each have finite order, yet xy has infinite order.

8.19 If H and K are subgroups of a group G, then HK denotes the set of all elements of G that can be written in the form hk, with $h \in H$ and $k \in K$. Find subgroups H and K of S_3 such that HK is not a subgroup of S_3.

8.20 (See Exercise 8.12.) Show that $Z(D_n)$ consists of e alone if n is odd, and $Z(D_n)$ contains exactly two elements if n is even.

8.21 Let p be a prime, and let H be a subgroup of S_p that contains a transposition and a p-cycle. Show that $H = S_p$.

EQUIVALENCE RELATIONS; COSETS

Many times in mathematics we run into the following kind of situation. We have a set S and we wish to identify certain elements of S with each other, i.e., to regard certain elements as being "essentially the same" even though they are different elements. This comes about when we are considering some relationship that may or may not hold between two elements of S, and we wish to "lump together" any two elements between which the relationship holds. For example, if we were considering the set of all triangles in the plane, we might want to regard as "the same" any two triangles that were congruent to each other.

Let's examine this general situation a bit more precisely. First of all, what do we mean by a "relationship" that may or may not hold between two elements of S? By a **relation** R on S we mean a set of ordered pairs of elements of S. If $s_1, s_2 \in S$, then s_1 is in the relationship R to s_2 if and only if the ordered pair (s_1, s_2) is one of the pairs in R. For example, if $S = \mathbb{Z}$ and R is the set of all pairs (m, n) where m and n are both even or both odd, then 2 is related to 6 by R, since 2 and 6 are both even $[(2, 6) \in R]$, but 2 is not related to 3, since 2 is even and 3 is odd $[(2, 3) \notin R]$. For convenience we usually express the fact that $(s_1, s_2) \in R$ by writing $s_1 R s_2$; thus in our example we have $2 R 6$ but not $2 R 3$. Intuitively, $s_1 R s_2$ just means that s_1 is related to s_2 by R.

Not every relation R on S is suitable for use in performing identifications. For example, we certainly want to identify any element s with itself, so if we aim to identify s_1 with s_2 if and only if $s_1 R s_2$, then we want R to have the property that $s R s$, for every $s \in S$. Similarly, if we are going to identify s_1 with s_2 then we want to identify s_2 with s_1, so we want R to have the property that $s_1 R s_2$ implies $s_2 R s_1$. Finally, if we identify s_1 and s_2 and also s_2 and s_3 then we want to identify s_1 and s_3, so we want it to be the case that whenever

$s_1 R s_2$ and $s_2 R s_3$ then $s_1 R s_3$. These considerations lead us to the definition of a special kind of relation on S called an *equivalence relation*.

DEFINITION A relation R on S is called an **equivalence relation** on S if R has the following three properties:

Reflexivity: For every $s \in S$, $s R s$;

Symmetry: For every s_1 and s_2 in S, if $s_1 R s_2$ then $s_2 R s_1$;

Transitivity: For every s_1, s_2, and s_3 in S, if $s_1 R s_2$ and $s_2 R s_3$ then $s_1 R s_3$.

Examples

1. Let S be any set and let R be the relation of equality on S, that is, $s_1 R s_2$ iff $s_1 = s_2$. Clearly R is an equivalence relation on S, and it is the smallest one in the sense that it relates two elements iff they are related by every equivalence relation on S.

2. Let $S = \mathbb{Z}$ and let $a R b$ iff $a < b$. R is not an equivalence relation, since the fact that $1 \not< 1$ means that reflexivity fails. Symmetry fails too; we have $1 < 2$, but $2 \not< 1$.

3. Let S be the set of all triangles in the plane and for two triangles $\triangle_1$ and $\triangle_2$, let $\triangle_1 R \triangle_2$ iff $\triangle_1$ is congruent to $\triangle_2$. It is easy to see that R is an equivalence relation.

4. Let $S = \mathbb{Z}$ and let $a R b$ iff a and b are both even or both odd. Again it is easy to see that R is an equivalence relation. R identifies all the even integers and identifies all the odd integers.

Notice that here $a R b$ iff $a \equiv b \pmod 2$. More generally, we obtain an equivalence relation on $\mathbb{Z}$ from any positive integer n, by defining $a R b$ iff $a \equiv b \pmod n$, that is, $n | (a - b)$. *Check:* We have $n | (a - a)$, so R is reflexive. If $n | (a - b)$, then $n | (b - a)$, so R is symmetric. If $n | (a - b)$ and $n | (b - c)$, then

$$n | [(a - b) + (b - c)],$$

that is, $n | (a - c)$, so R is transitive.

Before getting into some more sophisticated examples, we will see *how* an equivalence relation R on S accomplishes the identification of certain elements. We shall show that R breaks S up into a collection of mutually exclusive subsets (called *equivalence classes*).

For any $s \in S$, let $\bar{s}$ denote the subset of S consisting of all $x \in S$ such that $x R s$. In symbols,

$$\bar{s} = \{x \in S \mid x R s\}.$$

$\bar{s}$ is called the **equivalence class of s under R**.

THEOREM 9.1 Let R be an equivalence relation on S. Then every element of S is in exactly one equivalence class under R. That is, the equivalence classes partition S into a family of mutually disjoint nonempty subsets.

Conversely, given any partition of S into mutually disjoint nonempty subsets, there is an equivalence relation on S whose equivalence classes are precisely the subsets in the given partition of S.

PROOF. First half: Since for any $s \in S$ we have $s \in \bar{s}$ by virtue of the fact that R is reflexive, we see that every $s \in S$ is contained in some equivalence class (and that every equivalence class is nonempty). Now we will show that if two equivalence classes $\bar{s}_1$ and $\bar{s}_2$ are not disjoint, then they actually coincide. Suppose $x \in \bar{s}_1 \cap \bar{s}_2$. Then $x R s_1$ and $x R s_2$, so $s_1 R x$ and $x R s_2$, whence $s_1 R s_2$. Thus if y is any element of $\bar{s}_1$, we have $y R s_1$ and $s_1 R s_2$, so $y R s_2$, that is, $y \in \bar{s}_2$. This yields $\bar{s}_1 \subseteq \bar{s}_2$, and similarly we see that $\bar{s}_2 \subseteq \bar{s}_1$, so $\bar{s}_1 = \bar{s}_2$. We have established that any two *different* equivalence classes must be disjoint.

Second half: The second half of the theorem is not as important as the first half, and we leave its proof to the reader. □

It follows from the proof of the first half that $\bar{s}_1 = \bar{s}_2$ iff $\bar{s}_1 \cap \bar{s}_2 \neq \varnothing$. One also sees easily that $\bar{s}_1 = \bar{s}_2$ iff $s_1 R s_2$ (Exercise 9.3). In particular, if we are given an equivalence class $\bar{s}$, then $\bar{s}$ is also $\overline{s'}$, for any $s' \in \bar{s}$. Thus we can refer to an equivalence class by using any one of its elements. The elements of an equivalence class are sometimes called **representatives** of that class.

In Example 1 above, the equivalence classes consist of one element each. In Example 3, an equivalence class consists of all the triangles that are congruent to some one triangle. In Example 4, there are n equivalence classes —namely, those of $0, 1, \ldots, n-1$—since every integer is congruent to precisely one of $0, 1, \ldots, n-1 \pmod{n}$.

We have not yet indicated any connection between equivalence relations and group theory, but we don't have to look far to find one. For in checking that congruence $\bmod n$ (i.e., "differing by an element of $n\mathbb{Z}$") is an equivalence relation on $\mathbb{Z}$, we used the full strength of the fact that $n\mathbb{Z}$ is a *subgroup* of $(\mathbb{Z}, +)$. Specifically, we used the fact that $0 \in n\mathbb{Z}$ to check reflexivity, the fact that $n\mathbb{Z}$ is closed under inverses to check symmetry, and the fact that $n\mathbb{Z}$ is closed under addition to check transitivity.

We know that "congruence $\bmod n$" is a useful notion, and this suggests that it might be worthwhile to consider the general idea of two elements of a group G "differing" by an element of a given subgroup H. Since the analogue of the difference $x - y$ for arbitrary, possibly nonabelian groups is xy^{-1}, we consider the relation $\equiv_H$ given by

$$x \equiv_H y \quad \text{iff} \quad xy^{-1} \in H.$$

THEOREM 9.2 For any group G and any subgroup H of G, the relation $\equiv_H$ defined above is an equivalence relation.

PROOF. Just like the verification for congruence mod n, in different notation:

For any $x \in G$, $xx^{-1} = e \in H$, so $x \equiv_H x$, and $\equiv_H$ is reflexive. Next, if $x \equiv_H y$ then $xy^{-1} \in H$, so $(xy^{-1})^{-1} \in H$, since H is a subgroup; thus

$$\left(y^{-1}\right)^{-1} x^{-1} \in H,$$

that is, $yx^{-1} \in H$, and $y \equiv_H x$. Therefore, $\equiv_H$ is symmetric.

Finally, we claim that $\equiv_H$ is transitive. Suppose $x \equiv_H y$ and $y \equiv_H z$. Then $xy^{-1} \in H$ and $yz^{-1} \in H$, so

$$\left(xy^{-1}\right)\left(yz^{-1}\right) \in H \quad \text{(why?)},$$

that is, $xz^{-1} \in H$, and we have $x \equiv_H z$ as desired. $\square$

We are going to do some very interesting things with the equivalence classes of $\equiv_H$, and it will be useful to have another description of them.

DEFINITION If H is a subgroup of G, then by a **right coset of H in G** we mean a subset of G of the form Ha, where $a \in G$ and $Ha = \{ha \mid h \in H\}$.

Thus a coset of H is a "translate" of H by some element of G. For instance, if we think of $(\mathbb{R}, +)$ as the set of points on a line and consider the subgroup $\mathbb{Z}$, then the coset $\mathbb{Z} + \frac{1}{2}$ consists of the points we get by shifting all the integers one-half unit to the right:

[We have written $\mathbb{Z} + \frac{1}{2}$ rather than $\mathbb{Z}\frac{1}{2}$ because the operation in $(\mathbb{R}, +)$ is addition.] We will look at more examples after we make the connection between right cosets and $\equiv_H$ explicit.

THEOREM 9.3 Let H be a subgroup of G. For $a \in G$, let $\bar{a}$ denote the equivalence class of a under $\equiv_H$. Then

$$\bar{a} = Ha.$$

Thus the equivalence classes of $\equiv_H$ are precisely the right cosets of H.

PROOF. We show that if $x \in G$ then $x \in \bar{a}$ iff $x \in Ha$:

$$x \in \bar{a} \quad \text{iff} \quad x \equiv_H a \quad \text{iff} \quad xa^{-1} \in H \quad \text{iff} \quad x \in Ha. \quad \Box$$

COROLLARY 9.4 Let H be a subgroup of G, and let $a, b \in G$. Then $Ha = Hb$ iff $ab^{-1} \in H$.

PROOF. $Ha = Hb$ iff $\bar{a} = \bar{b}$ iff $a \equiv_H b$ iff $ab^{-1} \in H$. $\Box$

In particular, taking $b = e$, we see that $Ha = H$ iff $a \in H$.

Examples In the following examples, $\bar{a}$ always denotes equivalence classes with respect to $\equiv_H$.

1. Let $G = (\mathbb{Z}, +)$, and let $H = 6\mathbb{Z}$. The right cosets of H in G are:

$$6\mathbb{Z} = 6\mathbb{Z} + 0 = \{\ldots, -12, -6, 0, 6, 12, \ldots\} = 6\mathbb{Z} + 6 = 6\mathbb{Z} - 6 = \ldots = \bar{0} = \bar{6} = \overline{-6} = \ldots,$$

$$6\mathbb{Z} + 1 = \{\ldots, -11, -5, 1, 7, 13, \ldots\} = 6\mathbb{Z} + 7 = 6\mathbb{Z} - 5 = \ldots = \bar{1} = \bar{7} = \overline{-5} = \ldots,$$

$$6\mathbb{Z} + 2 = \{\ldots, -10, -4, 2, 8, 14, \ldots\} = 6\mathbb{Z} + 8 = 6\mathbb{Z} - 4 = \ldots = \bar{2} = \bar{8} = \overline{-4} = \ldots,$$

$$6\mathbb{Z} + 3 = \{\ldots, -9, -3, 3, 9, 15, \ldots\} = 6\mathbb{Z} + 9 = 6\mathbb{Z} - 3 = \ldots = \bar{3} = \bar{9} = \overline{-3} = \ldots,$$

$$6\mathbb{Z} + 4 = \{\ldots, -8, -2, 4, 10, 16, \ldots\} = 6\mathbb{Z} + 10 = 6\mathbb{Z} - 2 = \ldots = \bar{4} = \overline{10} = \overline{-2} = \ldots,$$

$$6\mathbb{Z} + 5 = \{\ldots, -7, -1, 5, 11, 17, \ldots\} = 6\mathbb{Z} + 11 = 6\mathbb{Z} - 1 = \ldots = \bar{5} = \overline{11} = \overline{-1} = \ldots.$$

2. Let $G = (\mathbb{Z}_{12}, \oplus)$, and let $H = \langle 4 \rangle = \{4, 8, 0\}$. The right cosets of H are:

$$H = H \oplus 0 = \{4, 8, 0\} = H \oplus 4 = H \oplus 8 = \bar{0} = \bar{4} = \bar{8},$$

$$H \oplus 1 = \{5, 9, 1\} = H \oplus 5 = H \oplus 9 = \bar{1} = \bar{5} = \bar{9},$$

$$H \oplus 2 = \{6, 10, 2\} = H \oplus 6 = H \oplus 10 = \bar{2} = \bar{6} = \overline{10},$$

$$H \oplus 3 = \{7, 11, 3\} = H \oplus 7 = H \oplus 11 = \bar{3} = \bar{7} = \overline{11}.$$

3. Let $G = \mathbb{Z}_3 \times \mathbb{Z}_2$, and let $H = \{(0,0), (0,1)\}$. H is a subgroup of G, and its right cosets are:

$$H = H + (0,0) = \{(0,0), (0,1)\} = H + (0,1) = \overline{(0,0)} = \overline{(0,1)},$$

$$H + (1,0) = \{(1,0), (1,1)\} = H + (1,1) = \overline{(1,0)} = \overline{(1,1)},$$

$$H + (2,0) = \{(2,0), (2,1)\} = H + (2,1) = \overline{(2,0)} = \overline{(2,1)}.$$

Note that two elements determine the same right coset of H iff they differ by something in H, i.e., they have the same first coordinate.

4. Let $G = S_3 = \{e, f, f^2, g, fg, f^2g\}$, where $o(f) = 3$, $o(g) = 2$, and $gf^i = f^{-i}g$. Let $H = \langle g \rangle = \{e, g\}$. Then we have:

$$H = He = \{e, g\} \quad = Hg \quad = \bar{e} \quad = \bar{g},$$

$$Hf = \{f, gf\} \quad = Hgf = \bar{f} \quad = \overline{gf},$$

$$Hf^2 = \{f^2, gf^2\} = Hgf^2 = \overline{f^2} = \overline{gf^2}.$$

5. Let $G = (\mathbb{Q}, +)$ and let $H = \mathbb{Z}$. We assert that the distinct right cosets of $\mathbb{Z}$ in $(\mathbb{Q}, +)$ are precisely the cosets $\mathbb{Z} + q$, for $0 \leqslant q < 1$.

First of all, if $q_1, q_2 \in \mathbb{Q}$, $0 \leqslant q_1 < 1$, $0 \leqslant q_2 < 1$, and $q_1 \neq q_2$, then $\mathbb{Z} + q_1 \neq \mathbb{Z} + q_2$, because q_1 and q_2 do not differ by something in $\mathbb{Z}$, that is, $q_1 - q_2$ is not an integer. Secondly, for any $r \in \mathbb{Q}$, the coset $\mathbb{Z} + r$ is the same as $\mathbb{Z} + q$ for some q, $0 \leqslant q < 1$, because there is such a q which differs from r by an integer.

Observe that, in this example, there are infinitely many cosets.

Having considered right cosets, it is natural to wonder about *left* ones. If H is a subgroup of G, then a **left coset** of H is of course a subset of G of the form $aH = \{ah | h \in H\}$, where $a \in G$. Arguments just like the ones we have gone through show that the left cosets of H are the equivalence classes under the equivalence relation $_H\equiv$ defined by

$$x \,_H\equiv y \quad \text{iff} \quad x^{-1}y \in H$$

(Exercise 9.13). The left cosets of H need not be the same as the right cosets.

Example Let's find the left cosets of the subgroup $H = \{e, g\}$ in S_3. We have:

$$H = eH = \{e, g\} = gH, \qquad \text{(Check it!)}$$

$$fH = \{f, fg\} = fgH,$$

$$f^2H = \{f^2, f^2g\} = f^2gH.$$

Observe that H is both a left coset and a right coset, but no other left coset is a right coset. We will later single out for special attention those subgroups with the property that right cosets and left cosets are the same thing.

EXERCISES

9.1 Determine which of the following relations R on $\mathbb{Z}$ are equivalence relations.

  a) $a\,R\,b$ iff $a - b > 0$
  b) $a\,R\,b$ iff $|a| = |b|$
  c) $a\,R\,b$ iff $ab > 0$
  d) $a\,R\,b$ iff $|a - b| < 1$

9.2 Let $S = \{a,b,c\}$ and let $R = \{(a,a),(b,b),(c,c),(a,b),(b,a),(c,b)\}$. Which of the properties of an equivalence relation hold for R?

9.3 Let R be an equivalence relation on S. Show that for any $s_1, s_2 \in S$, we have
$$\bar{s}_1 = \bar{s}_2 \quad \text{iff} \quad s_1 R s_2.$$

9.4 Let $G = Q_8$. Find the right cosets of H in G for:
a) $H = \langle J \rangle$.
b) $H = \langle -I \rangle$.

9.5 Let $G = D_4$ and let $H = \{e, f^2 g\}$. Find the right cosets of H in G, and the left cosets.

9.6 Find the right cosets of the subgroup $H = \{(0,0),(1,0),(2,0)\}$ in $\mathbb{Z}_3 \times \mathbb{Z}_2$.

9.7 Find the right cosets of the subgroup $H = \{(0,0),(0,2)\}$ in $\mathbb{Z}_4 \times \mathbb{Z}_4$.

9.8 Find the right cosets of the subgroup $H = \langle (1,1) \rangle$ in $\mathbb{Z}_2 \times \mathbb{Z}_4$.

9.9 Let $X = \{1,2,3,4\}$ and let $Y = \{1,2\}$. Let G be the group $(P(X), \triangle)$, H the subgroup $(P(Y), \triangle)$. Find the right cosets of H in G.

9.10 For sets A and B, define $A R B$ iff there exists a one-to-one mapping from A onto B. Show that R is an equivalence relation on the class of all sets.

9.11 Let G be a group and for elements $a, b \in G$ define $a R b$ iff there exists an element $x \in G$ such that $a = xbx^{-1}$. Show that R is an equivalence relation on G.

9.12 Let G be a group and for $a, b \in G$ define $a R b$ iff $ab = ba$. Must R be an equivalence relation on G? If so, prove it; if not, can you indicate for which groups R is an equivalence relation?

9.13 Let H be a subgroup of a group G and define $_H\equiv$ on G by $x \;_H\equiv\; y$ iff $x^{-1}y \in H$.
a) Show that $_H\equiv$ is an equivalence relation on G.
b) Show that the equivalence classes under $_H\equiv$ are the left cosets of H in G.
c) Show that for $a, b \in G$, $aH = bH$ iff $a^{-1}b \in H$.

9.14 Prove the second half of Theorem 9.1.

COUNTING THE ELEMENTS OF A FINITE GROUP

In this section we shall consider two different ways of counting the elements of a finite group G. One way will lead us to a classic result known as *Lagrange's Theorem*; the other will yield a powerful tool called the *class equation* of G. In both cases, the plan will be to use an equivalence relation to split G up into disjoint subsets, and then to count the elements of G by counting those in each subset separately and adding the answers.

We begin with

THEOREM 10.1 (Lagrange's Theorem) Let G be a finite group and let H be a subgroup of G. Then $|H|$ divides $|G|$.

Lagrange's Theorem has an obvious significance in that it says a lot about what possibilities there are for subgroups of a given finite group. This information can be used to derive a variety of interesting results with little or no effort; but first let's prove the theorem itself. This too is not hard, because the groundwork was laid in the previous section. Notice, incidentally, that we already know the theorem is true if G happens to be cyclic.

We do need the following easy preliminary result:

LEMMA 10.2 Let G be any group (not necessarily finite), and let H be a subgroup of G. Let Ha and Hb be right cosets of H in G. Then there is a one-to-one correspondence between the elements of Ha and those of Hb.

PROOF. Define a function $f: Ha \to Hb$ by declaring $f(ha) = hb$ for every $h \in H$. Then f is onto because every element of Hb has the form hb for some $h \in H$; and f is one-to-one since if $f(h_1 a) = f(h_2 a)$—that is, if $h_1 b = h_2 b$—then $h_1 = h_2$ by right cancellation, so $h_1 a = h_2 a$. $\square$

The significance of this result is that any two right cosets of H in G have the same number of elements. Thus, for example, if one right coset has sixteen elements, then every right coset has sixteen elements. If one right coset is infinite, then every right coset is infinite, and has the "same infinite number" of elements, because of the one-to-one correspondence exhibited above. In general, for any two sets S and T, we say that S and T have the same **cardinality**, and we write $|S| = |T|$, if there exists a one-to-one correspondence between the elements of S and those of T. Thus we can express the result of the lemma by saying that any two right cosets of H have the same cardinality.

PROOF OF LAGRANGE'S THEOREM. Let G, H be as in the statement of the theorem. The idea of the proof is that we can split G up into a finite number of mutually disjoint subsets, each having $|H|$ elements. Thus $|G|$ is $|H|$ times the number of subsets.

Let $\equiv_H$ denote the equivalence relation given by $a \equiv_H b$ iff $ab^{-1} \in H$. By Theorem 9.1, the equivalence classes under $\equiv_H$ partition G into a collection of mutually disjoint nonempty subsets, and by Theorem 9.3, these equivalence classes are just the right cosets of H. Since G is finite, only finitely many distinct cosets can fit into G, so we have

$$G = Ha_1 \cup Ha_2 \cup \cdots \cup Ha_k$$

for some integer k and elements $a_1, a_2, \ldots, a_k$ in G. Now, by Lemma 10.2, all the cosets have the same number of elements, namely $|H|$, since H is one of the cosets. Thus, counting the elements on both sides of the last equation, we get

$$|G| = |H| + |H| + \cdots + |H|,$$

where there are k terms on the right. Thus $|G| = k \cdot |H|$, and we have shown that $|H|$ divides $|G|$. $\square$

If G is any group (not necessarily finite) and H is any subgroup, then the number of distinct right cosets of H in G is called the **index** of H in G. We denote this number by $[G:H]$. Thus Lagrange's Theorem tells us that if G is finite, we have

$$|G| = [G:H] \cdot |H|.$$

Examples

1. Let G be Klein's 4-group, that is, $G = \{e, a, b, c\}$, with $a^2 = b^2 = c^2 = e$, $ab = ba = c$, $ac = ca = b$, and $bc = cb = a$. Then if $H = \langle a \rangle$, $|H| = 2$, and since $|G| = 4$, we get $4 = [G:H] \cdot 2$, so $[G:H] = 2$. In fact the right cosets of H are $H = \{e, a\}$ and $Hb = \{b, c\}$.

2. Let $G = (\mathbb{Z}_{12}, \oplus)$ and let $H = \langle 4 \rangle$. Then $|H| = 3$, and the right cosets of H in G are $H = \{4, 8, 0\}$, $H \oplus 1 = \{5, 9, 1\}$, $H \oplus 2 = \{6, 10, 2\}$, and $H \oplus 3 = \{7, 11, 3\}$. We have $|G| = [G:H] \cdot |H|$, that is, $12 = 4 \cdot 3$.

3. Let $G = S_3$ and let $H = \langle g \rangle$. Then $|H| = 2$, so $|G| = [G:H] \cdot |H|$ says that $6 = [G:H] \cdot 2$, or $[G:H] = 3$. As we saw in Section 9, the distinct right cosets of H in G are H, Hf, and Hf^2.

4. For an easy example in which G is infinite, let $G = (\mathbb{Z}, +)$ and let $H = \langle 2 \rangle = 2\mathbb{Z}$. The right cosets of H are H and $H + 1$, so $[G:H] = 2$.

5. It is possible for $[G:H]$ to be infinite. For instance, let $G = (\mathbb{Q}, +)$ and $H = \mathbb{Z}$. We saw in Section 9 that there are infinitely many right cosets of H in G.

We have defined $[G:H]$ to be the number of right cosets of H in G. It is natural to wonder what the number of left cosets is. Since the left cosets need not be the same as the right ones, it comes as a pleasant surprise that the *number* of left cosets is always the same as the *number* of right ones (even if G is infinite).

THEOREM 10.3 Let H be a subgroup of G. Then the number of left cosets of H in G is $[G:H]$.

PROOF. Let S be the set of all right cosets of H, and let T be the set of all left cosets. To prove the theorem, we will show that there is a one-to-one function from S onto T.

We want to define $f: S \to T$ by $f(Ha) = a^{-1}H$, *for all* $a \in G$. The one thing we have to worry about is whether this assigns a unique value to each right coset. In other words, if $Hc = Hd$, then we have mapped this coset to $c^{-1}H$, but also to $d^{-1}H$. Does $c^{-1}H = d^{-1}H$? It does, because if $Hc = Hd$, then $cd^{-1} \in H$, that is, $(c^{-1})^{-1}d^{-1} \in H$. By Exercise 9.13, this says $c^{-1}H = d^{-1}H$.

Thus we have assigned a unique value in T to each element of S. Our function f is one-to-one by the reverse of the argument we just gave. That is, if $f(Hc) = f(Hd)$, i.e., $c^{-1}H = d^{-1}H$, then $(c^{-1})^{-1}d^{-1} \in H$, $cd^{-1} \in H$, and $Hc = Hd$. Finally, f is onto since if cH is any element of T, we have $Hc^{-1} \in S$, and $f(Hc^{-1}) = (c^{-1})^{-1}H = cH$. $\square$

A word more about this proof. Instead of attempting to achieve $f(Ha) = a^{-1}H$ for *every* $a \in G$, we might have tried choosing one particular representative x for each right coset Hx, and defining $f(Hx) = x^{-1}H$ *just for that special representative*. This approach is not as satisfactory as the one we adopted. It involves arbitrary choices, hence is not as "natural," and it also binds us to the particular representatives chosen. For instance, in checking that f is onto, we took an arbitrary $cH \in T$ and said that $f(Hc^{-1}) = cH$; if we had used fixed representatives for all the cosets, we would have had to worry about whether c^{-1} was the fixed representative for Hc^{-1}, in order to get $f(Hc^{-1}) = cH$.

We will now look at several results that demonstrate the importance of Lagrange's Theorem.

THEOREM 10.4 Let G be a finite group and let $x \in G$. Then $o(x)$ divides $|G|$. Consequently, $x^{|G|} = e$ for every $x \in G$.

PROOF. If we are going to use Lagrange's Theorem to show that $o(x)$ divides $|G|$, then obviously what we want to do is to find ourselves a subgroup H of G such that $|H| = o(x)$. Take $H = \langle x \rangle$; then indeed $|H| = o(x)$, so $o(x)$ does divide $|G|$.

From this it is immediate that $x^{|G|} = e$. $\square$

THEOREM 10.5 Let G be a group and assume that $|G|$ is a prime. Then G is cyclic. Moreover, any element of G other than e is a generator for G.

PROOF. If $|G|$ is a prime p, then $|G| > 1$, so G is not trivial and there is an element $x \neq e$ in G. Then by Lagrange's Theorem, $|\langle x \rangle|$ divides p; since $|\langle x \rangle| \neq 1$ this forces $|\langle x \rangle| = p$, because p is prime. Hence $\langle x \rangle$ contains the same number of elements as G, so $\langle x \rangle = G$, and any nonidentity element of G is a generator for G. $\square$

To illustrate the use of Theorems 10.4 and 10.5, we will show that every group G of order $\leqslant 5$ is abelian, thus verifying a claim we made in Section 8. If $|G| = 1, 2, 3$, or 5, then by Theorem 10.5, G is cyclic, hence abelian. If $|G| = 4$, then by Theorem 10.4, every nonidentity element of G has order either 2 or 4. If there is an element of order 4, then G is cyclic; otherwise we have $x^2 = e$ for every $x \in G$, so G is abelian by Exercise 3.11.

Actually, the appeal to Exercise 3.11 can be avoided by the following direct argument. Suppose that $G = \{e_G, A, B, C\}$ and A, B, C are each of order 2. Then AB can't be e_G, because the inverse of A is A, not B; and AB can't be A or B, since neither A nor B is e_G. Thus AB must be C, and, similarly, BA must be C. The same argument shows that $BC = CB = A$ and $AC = CA = B$, so G is abelian. Observe that this reasoning reveals more than the fact that G is abelian; it shows that G is essentially the same as $V = \{e, a, b, c\}$, with A, B, C playing the roles of a, b, c. Putting it another way, if we take the multiplication table for G and replace e_G, A, B, C everywhere in it by e, a, b, c, respectively, then we get the multiplication table for V.

It follows that there are essentially only two possibilities for the structure of a group G of order 4. If G is cyclic, say $G = \langle x \rangle$, then the operation in G is just like that in $(\mathbb{Z}_4, \oplus)$, with e, x, x^2, x^3 playing the roles of $0, 1, 2, 3$. If G is not cyclic, then, as above, G is essentially the same as V. Along the same lines, there is only *one* possibility for the structure of a group of order 2, 3, or 5; the group is cyclic, hence essentially the same as $(\mathbb{Z}_2, \oplus)$, $(\mathbb{Z}_3, \oplus)$, or $(\mathbb{Z}_5, \oplus)$. We will make the notion of two groups being "essentially the same" precise in Section 12.

Right now we want to give an application of Lagrange's Theorem in number theory. To do this we introduce, for any prime p, the group $(\mathbb{Z}_p - \{0\}, \odot)$ of integers relatively prime to p under multiplication modulo p. The elements of $(\mathbb{Z}_p - \{0\}, \odot)$ are $1, 2, \ldots, p-1$; the operation is given by $m \odot n = \overline{mn}$, where $\overline{}$ denotes remainders modulo p. This does define an operation on $\mathbb{Z}_p - \{0\}$, since if $m, n \in \mathbb{Z}_p - \{0\}$, then neither m nor n is divisible by p, so their product mn is not divisible by p, and $\overline{mn}$ is not 0, i.e., is in $\mathbb{Z}_p - \{0\}$. We leave it as an exercise (10.17) to check that this operation is associative.

Clearly the identity element is 1. Finally, to find an inverse for $m \in \mathbb{Z}_p - \{0\}$, we need $n \in \mathbb{Z}_p - \{0\}$ such that $\overline{mn} = 1$; in other words, we need n such that $mn = 1 + kp$ for some integer k. This is the same as asking for an n and a k such that $mn + kp = 1$. The existence of n and k is guaranteed by Theorem 4.2, because $(m, p) = 1$.

Thus $(\mathbb{Z}_p - \{0\}, \odot)$ is a group. Using it, we can easily prove the following classic result of number theory, which was stated by Pierre de Fermat in 1640. The first published proof was given by Leonhard Euler in 1736.

THEOREM 10.6 (Fermat's Theorem) Let p be a prime and suppose a is an integer such that p does not divide a. Then $a^{p-1} \equiv 1 \pmod{p}$.

As an example, $37^{22} \equiv 1 \pmod{23}$.

PROOF. If a is one of $1, 2, 3, \ldots, p-1$, then by applying Theorem 10.4 to the group $G = (\mathbb{Z}_p - \{0\}, \odot)$, we see that $a^{p-1} = e$ in G, since $|G| = p - 1$. This says that $a^{p-1} \equiv 1 \pmod{p}$, as desired.

Now if a is any integer not divisible by p, then we can write $a = qp + r$, where $0 < r < p$; then $a^{p-1} \equiv r^{p-1} \pmod{p}$ (why?), hence $a^{p-1} \equiv 1$ by what we have already shown. $\square$

If you are familiar with number theory, you will recall that there is a generalization of Fermat's Theorem, called **Euler's Theorem**. This says that if m is *any* positive integer, not necessarily prime, and if a is an integer such that $(a, m) = 1$, then $a^{\phi(m)} \equiv 1 \pmod{m}$, where $\phi(m)$ denotes the number of non-negative integers less than m that are relatively prime to m. This generalization can be proved by essentially the same technique we used above for the special case. One considers the group of integers in the set $\{0, 1, 2, \ldots, m-1\}$ that are relatively prime to m, under multiplication mod m. Since the order of this group is $\phi(m)$, the result follows. If you know about this theorem, you might be interested in checking through the details of the proof.

We conclude our discussion of Lagrange's Theorem by mentioning that the theorem does not have a complete converse. That is, if G is a finite group and n is an integer which divides $|G|$, then it does *not* follow that G has a subgroup of order n. (You will recall that if G is *cyclic*, then it *does* follow.) The usual counterexample is the group $G = A_4$, the alternating group of degree 4. We know that $|G| = 4!/2 = 12$, but it turns out that although 6 divides 12, A_4 has no subgroup of order 6. (See Exercise 11.20.)

We now turn to our second major objective for this section: the class equation.[†] We need some preliminary ideas.

In Exercise 5.23 we introduced the notion of the **centralizer** of an element. If G is a group and $y \in G$, then the centralizer of y, denoted by $Z(y)$, is the set of all elements that commute with y:

$$Z(y) = \{ x \in G \mid xy = yx \}.$$

The point of Exercise 5.23 was that $Z(y)$ is always a subgroup of G. It is clear that $Z(G) \subseteq Z(y)$ (an element that commutes with everything certainly commutes with y!), but in general $Z(y)$ may be larger than $Z(G)$.

Examples

1. If G is abelian, then $Z(y) = G$ for every $y \in G$.

2. Let $G = S_3 = \{ e, f, f^2, g, fg, f^2g \}$. Then $Z(f) = \{ e, f, f^2 \}$, since we can verify by direct calculation that g, fg, and f^2g do not commute with f. Likewise, $Z(g) = \{ e, g \}$, since f, f^2, fg, and f^2g do not commute with g. In this case, $Z(G) = \{ e \}$, the trivial subgroup, so both $Z(f)$ and $Z(g)$ are larger than $Z(G)$.

3. Let $G = D_4 = \{ e, f, f^2, f^3, g, fg, f^2g, f^3g \}$, where $o(f) = 4$, $o(g) = 2$, and $gf^i = f^{-i}g$ for all i. Here $Z(g) = \{ e, f^2, g, f^2g \}$ (verify!), so $Z(g)$ is larger than $\langle g \rangle$, but smaller than G.

Now let G be a group. Define a relation R on G by

$$a \, R \, b \text{ iff } a \text{ is conjugate to } b,$$

that is, $a \, R \, b$ iff there exists some $x \in G$ such that $a = xbx^{-1}$.

LEMMA 10.7 The above relation R is an equivalence relation on G.

[†]The remaining material of this section is needed for Exercises 10.20–10.25 and for Section 15, but can otherwise be omitted without loss of continuity.

PROOF. For any $a \in G$, we have $a = eae^{-1}$, so $a R a$, and R is reflexive. If $a R b$, then $a = xbx^{-1}$ for some x, whence

$$b = x^{-1}ax = x^{-1}a(x^{-1})^{-1},$$

so $b R a$, and R is symmetric. Finally, if $a R b$ and $b R c$, then we have $a = xbx^{-1}$ and $b = ycy^{-1}$ for some $x, y \in G$. Thus

$$a = x(ycy^{-1})x^{-1} = (xy)c(xy)^{-1},$$

so $a R c$, and R is transitive. $\square$

The equivalence class $\bar{a}$ of an element $a \in G$ under R is called the **conjugacy class** of a, and consists of all the conjugates of a. That is, $\bar{a} = \{ xax^{-1} | x \in G \}$.

Example In $S_3 = \{ e, f, f^2, g, fg, f^2g \}$, we have $\bar{e} = \{ e \}$. To find $\bar{f}$, notice that f^2 is a conjugate of f, because $f^2 = gfg^{-1}$. Since every conjugate of f must have the same order as f, namely 3, none of g, fg, f^2g is conjugate to f, and therefore $\bar{f} = \{ f, f^2 \}$. As for $\bar{g}$, the equations $fg = f^2g(f^2)^{-1}$ and $f^2g = fgf^{-1}$ show that both fg and f^2g are conjugates of g, and we see that $\bar{g} = \{ g, fg, f^2g \}$.

Thus R breaks S_3 up as

$$S_3 = \{ e \} \cup \{ f, f^2 \} \cup \{ g, fg, f^2g \}.$$

Our general aim is to count an arbitrary finite group G by counting each conjugacy class separately, so we need to find an expression for the number of conjugates of an element a. In the example of S_3, the element f has two conjugates, and two also happens to be the index of the subgroup $Z(f) = \{ e, f, f^2 \}$ in S_3. The element g has three conjugates, and again this number coincides with the index of $Z(g)$. The element e has only one conjugate, and, sure enough, $[S_3 : Z(e)] = 1$.

LEMMA 10.8 Let G be a finite group, and let $a \in G$. Then the number of distinct conjugates of a in G is $[G : Z(a)]$.

PROOF. Let $x, y \in G$. Then $xax^{-1} = yay^{-1}$ iff $ax^{-1}y = x^{-1}ya$ iff $x^{-1}y \in Z(a)$ iff $x \underset{Z(a)}{\equiv} y$ iff $xZ(a) = yZ(a)$. In other words, x and y determine the same conjugate of a iff they determine the same left coset of $Z(a)$, so the mapping

$$xax^{-1} \to xZ(a)$$

establishes a one-to-one correspondence between the conjugates of a and the left cosets of $Z(a)$ in G. By Theorem 10.3, the number of left cosets of $Z(a)$ is $[G : Z(a)]$, so the number of conjugates of a is $[G : Z(a)]$. $\square$

THEOREM 10.9 (Class equation) Let G be a finite group, and let $\{ a_1, \ldots, a_k \}$ consist of one element from each conjugacy class containing at least two elements. Then

$$|G| = |Z(G)| + [G:Z(a_1)] + [G:Z(a_2)] + \cdots + [G:Z(a_k)].$$

PROOF. Let $a_{k+1}, \ldots, a_{k+s}$ be the elements of the conjugacy classes containing only one element each. Then we know that G is the disjoint union

$$G = \bar{a}_1 \cup \bar{a}_2 \cup \cdots \cup \bar{a}_k \cup \{a_{k+1}\} \cup \cdots \cup \{a_{k+s}\}.$$

By Lemma 10.8, $\bar{a}_j$ contains exactly $[G:Z(a_j)]$ many elements, for $j = 1, 2, \ldots, k$. Thus

$$|G| = [G:Z(a_1)] + [G:Z(a_2)] + \cdots + [G:Z(a_k)] + s.$$

Also by Lemma 10.8, we see that for any $a \in G$, $\bar{a}$ consists of a alone $\Leftrightarrow$ $[G:Z(a)] = 1 \Leftrightarrow Z(a) = G \Leftrightarrow a \in Z(G)$, and therefore $a_{k+1}, \ldots, a_{k+s}$ are precisely the elements of $Z(G)$. Hence $s = |Z(G)|$, and we are done. $\square$

Perhaps the most noticeable difference between the situation here and that in the proof of Lagrange's Theorem is that here it need not be the case that all the equivalence classes have the same number of elements. For instance, we have already seen that

$$S_3 = \bar{e} \cup \bar{f} \cup \bar{g},$$

so the class equation of S_3 is

$$6 = 1 + 2 + 3.$$

The class equation has a number of striking applications. Two are presented in Exercises 10.24 and 10.25, and we will see another one in Section 15.

EXERCISES

10.1 Let $G = Q_8$. Find $[G:H]$ for $H = \langle -I \rangle$, $H = \langle K \rangle$, and $H = \langle -L \rangle$.

10.2 a) In $G = (\mathbb{Z}_{48}, \oplus)$, find $[G:H]$ for $H = \langle 32 \rangle$.

  b) In $G = (\mathbb{Z}_{54}, \oplus)$, find $[G:H]$ for $H = \langle 24 \rangle$.

  c) In $G = (\mathbb{Z}_{112}, \oplus)$, find $[G:H]$ for $H = \langle 100 \rangle$.

10.3 Let $G = \mathbb{Z}_6 \times \mathbb{Z}_4$, and find $[G:H]$ for:

  a) $H = \{0\} \times \mathbb{Z}_4$.

  b) $H = \langle 2 \rangle \times \langle 2 \rangle$.

10.4 Let $X = \{1, 2, 3, 4, 5\}$ and let $Y = \{1, 2, 3\}$. Let $G = (P(X), \triangle)$, $H = (P(Y), \triangle)$. Find $[G:H]$.

10.5 Let G be a group of order 8 that is not cyclic. Show that $a^4 = e$ for every $a \in G$.

10.6 Let G be a group and let H, K be subgroups of G such that $|H| = 12$ and $|K| = 5$. Prove that $H \cap K = \{e\}$.

10.7 Let p and q be two prime numbers, and let G be a group of order pq. Show that every proper subgroup of G is cyclic.

10.8 Let G be a group of order p^2, where p is a prime. Show that G must have a subgroup of order p.

10.9 Let H be a subgroup of G, and let Ha be a right coset of H other than H itself. Show that Ha is *not* a subgroup of G.

10.10 Provide another version of the proof of Lagrange's Theorem, as follows: List the elements of H: $h_1, h_2, \ldots, h_n$. If these elements do not exhaust G, pick a_1 not in the list and take the elements $h_1 a_1, h_2 a_1, \ldots, h_n a_1$. Show that the $2n$ elements written down so far are all distinct. If they do not exhaust G, pick an element a_2 not listed so far and write down $h_1 a_2, h_2 a_2, \ldots, h_n a_2$. Show that the $3n$ elements written down so far are all distinct. Continuing in this way, finish the proof.

10.11 a) Show that if H is a subgroup of G, then all the left cosets of H in G have the same number of elements.

b) Show that any right coset has the same number of elements as any left coset.

10.12 Show that if H is a subgroup of a *finite* group G, then the result of Theorem 10.3 can be established by redoing the proof of Lagrange's Theorem using left cosets instead of right ones.

10.13 In the proof of Theorem 10.3, try defining $f(Ha) = aH$, for all $a \in G$. Show that this may not give a well-defined function, because Ha may equal Hb without aH equaling bH. (*Suggestion:* Let $G = S_3$.)

10.14 Show that there are essentially only two groups of order 6, as follows.

a) If $|G| = 6$ and G contains an element of order 6, then G is cyclic. Why?

b) If G is not cyclic, then all elements of G have order 1, 2, or 3. Why? Show in fact that there must be an element of order 3. Call it a.

c) Let b be an element of G that is not in $\langle a \rangle$. Show that $e, a, a^2, b, ab, a^2 b$ are all the distinct elements of G.

d) Show that $o(b) = 2$. Since b was chosen arbitrarily, it follows that also $o(ab) = 2$ and $o(a^2 b) = 2$.

e) Show that $ba = a^2 b$ and $ba^2 = ab$.

The above steps show that either G is cyclic, in which case the multiplication in G is like that in $(\mathbb{Z}_6, \oplus)$, or G is not cyclic, in which case the multiplication in G is like that in S_3, with a and b playing the roles of the elements f and g.

10.15 Let G be a finite group, and let H be a subgroup of G. Let K be a subgroup of H. Prove that $[G:K] = [G:H][H:K]$.

10.16 Let G be an abelian group such that $|G|$ is an odd integer. Show that the product of all the elements in G is e.

10.17 Show that the multiplication in $\mathbb{Z}_p - \{0\}$ is associative.

10.18 Carry out the proof of Euler's Theorem suggested in the text.

10.19 Show that the set of even integers has the same cardinality as the set of odd integers.

10.20 Let G be a group, and suppose there is $g \in G$ such that $Z(g) = Z(G)$. Show that G is abelian.

10.21 Find the conjugacy classes in Q_8, and write down the class equation for Q_8.

10.22 Find the conjugacy classes in D_4, and write down the class equation for D_4.

10.23 Let G be a finite group. Show that $[G:Z(G)]$ cannot be a prime.

10.24 Let p be a prime number, and let G be a group such that $|G| = p^n$, where n is a positive integer. Use the class equation to show that $Z(G)$ is not trivial, that is, it contains more than one element. (*Hint:* Show that $|Z(G)|$ is divisible by p.) This result is a good example of what you can get by counting.

10.25 Let p be a prime number, and let G be a group such that $|G| = p^2$. Prove that G is abelian.

10.26 *Wilson's Theorem* (Another application of the group $(\mathbb{Z}_p - \{0\}, \odot)$ to number theory):

a) Show that if p is a prime, then $(p-1)! \equiv -1 \pmod{p}$. [*Hint:* Consider which elements of $(\mathbb{Z}_p - \{0\}, \odot)$ are their own inverses.] This result was stated by John Wilson (1741–1793). It also seems to have been known to Leibniz in the late 1600s. The first published proof of the theorem was given by Lagrange in 1770.

b) Prove the converse of the result in (a): Show that if $n > 2$ and $(n-1)! \equiv -1 \pmod{n}$, then n is prime.

NORMAL SUBGROUPS

At the end of Section 9 we pointed out that we would want to return to an investigation of those subgroups of a given group G which have the property that right cosets are the same thing as left cosets. We will now take up this question. The main reason why such subgroups are important is that it is possible to turn the set of right cosets (which is, in this case, the same thing as the set of left cosets) into a group in a "natural" way, i.e., in such a way that the multiplication on the cosets is inherited from the given multiplication on G.

We will begin by giving an alternative description of the special subgroups we have in mind.

DEFINITION Let H be a subgroup of G. Then we say that H is a **normal** subgroup if for every $h \in H$ and $g \in G$, ghg^{-1} is in H.

This definition can be expressed in terms of subsets by letting, for $g \in G$, $gHg^{-1} = \{ ghg^{-1} | h \in H \}$, and saying that H is normal if and only if $gHg^{-1} \subseteq H$ for every $g \in G$.

It should be observed that the definition does *not* say that $ghg^{-1} = h$ for every $g \in G$ and $h \in H$; rather it says that ghg^{-1} will be *some* element of H, not necessarily h. In fact $ghg^{-1} = h$ iff $gh = hg$, that is, iff g and h commute with each other.

THEOREM 11.1 Let H be a subgroup of G. Then the following are equivalent:

  i) H is normal;

 ii) $gHg^{-1} = H$ for every $g \in G$;

iii) $gH = Hg$ for every $g \in G$, i.e., the left cosets are the same as the right cosets.

Remark. Condition (iii) does *not* say that for every $h \in H$ we have $gh = hg$. Condition (iii) implies that gh will be $h'g$ for some $h' \in H$, but it is not necessary for h' to be h.

PROOF. (i) $\Leftrightarrow$ (ii): Clearly, (ii) implies (i). Now assume (i), i.e., assume that $gHg^{-1} \subseteq H$ for *every* $g \in G$. Then for any g, $g^{-1}H(g^{-1})^{-1} \subseteq H$. Thus $g^{-1}H \subseteq Hg^{-1}$, and $H \subseteq gHg^{-1}$. This, with the assumed $gHg^{-1} \subseteq H$, gives $gHg^{-1} = H$. Therefore, (i) implies (ii).

(ii) $\Leftrightarrow$ (iii): This is immediate, since (ii) and (iii) are obtained from each other by multiplying on the right by g or g^{-1}. $\square$

Examples

1. We saw in Section 9 that the subgroup $\langle g \rangle$ of $S_3 = \{e,f,f^2,g,fg,f^2g\}$ is not normal, since its right cosets do not coincide with its left cosets. However, the subgroup $\langle f \rangle$ $(=A_3)$ *is* a normal subgroup of S_3. Its right cosets are $A_3 = \{e,f,f^2\}$ and $A_3g = \{g,fg,f^2g\}$, and its left cosets are A_3 and $gA_3 = \{g,gf,gf^2\}$, the same two subsets.

Observe how $yay^{-1} \in A_3$ for every $a \in A_3$: for example, if y is f, then $fef^{-1} = e$, $fff^{-1} = f$, and $ff^2f^{-1} = f^2$; if y is g, then $geg^{-1} = e$, $gfg^{-1} = f^2gg^{-1} = f^2$, and $gf^2g^{-1} = fgg^{-1} = f$.

2. Let G and H be groups, and consider $G \times H$. The subgroup $\{e_G\} \times H$ is normal, because if $(g,h) \in G \times H$ and $(e_G,x) \in \{e_G\} \times H$, then

$$(g,h)(e_G,x)(g,h)^{-1} = (ge_Gg^{-1}, hxh^{-1}) = (e_G, hxh^{-1}) \in \{e_G\} \times H.$$

Likewise, the subgroup $G \times \{e_H\}$ is normal.

It is worthwhile to develop some criteria that can be used to guarantee that certain subgroups are normal without referring back to the definition of normality. We shall give three such criteria, one involving the idea of elements commuting with each other, and two involving counting arguments.

We have observed that the center $Z(G)$ of a group G is always a subgroup of G. We now show that $Z(G)$ is always a *normal* subgroup. In fact, more is true:

THEOREM 11.2 Let G be a group. Then any subgroup of $Z(G)$ is a normal subgroup of G.

PROOF. Let H be any subgroup of $Z(G)$. One way to show that H is normal in G is to show that $ghg^{-1} \in H$ for every $g \in G$ and $h \in H$. But since $h \in Z(G)$, we have $ghg^{-1} = hgg^{-1} = h \in H$. $\square$

COROLLARY If G is abelian, then every subgroup of G is normal.

PROOF. If G is abelian, then $Z(G) = G$, so every subgroup of G is a subgroup of $Z(G)$, hence normal by the theorem. ☐

THEOREM 11.3 Let H be a subgroup of G such that $[G:H] = 2$. Then H is normal in G.

PROOF. We will show that for any $g \in G$, $gH = Hg$.
   Since there are exactly two right cosets of H in G, the right cosets must be H and $G - H$. Likewise, the two left cosets must be H and $G - H$. Thus the left cosets are the same as the right cosets, and for any $g \in G$, the left coset containing g is also the right coset containing g, that is, $gH = Hg$. ☐

Examples

1. We have seen that if G is abelian then every subgroup of G is normal. A very natural question is whether the converse is true: If G is a group such that every subgroup of G is normal, must G be abelian? The answer is no, as is shown by taking $G = Q_8$. Our primary reason for introducing the group Q_8 in the first place was to have it available as an example here.
   The subgroups of Q_8 are $\langle I \rangle$, $\langle -I \rangle$, $\langle J \rangle$, $\langle K \rangle$, $\langle L \rangle$, and Q_8. (Recall that $Q_8 = \{I, -I, J, -J, K, -K, L, -L\}$.) Now Q_8 is normal in Q_8 since any group is obviously a normal subgroup of itself; $\langle J \rangle$, $\langle K \rangle$, and $\langle L \rangle$ are normal because they each have index 2 in Q_8; $\langle I \rangle$ and $\langle -I \rangle$ are normal because they are contained in $Z(Q_8)$.
   Thus every subgroup of Q_8 is normal, but Q_8 is not abelian. Nonabelian groups with the property that every subgroup is normal are called **Hamiltonian**, after the Irish mathematician Sir William Rowan Hamilton. It can be shown that every Hamiltonian group must have a subgroup which "looks just like" Q_8, so Q_8 is the simplest Hamiltonian group.

2. We have seen that A_3 is a normal subgroup of S_3. More generally, A_n is a normal subgroup of S_n for any $n \geqslant 2$, because $|A_n| = n!/2 = |S_n|/2$, so $[S_n : A_n] = 2$.

   Our last criterion for normality will follow as an immediate corollary of

THEOREM 11.4 Let G be a group, H a subgroup of G, and $g \in G$. Then gHg^{-1} is a subgroup of G, with the same number of elements as H.

PROOF. To see that gHg^{-1} is a subgroup, note that gHg^{-1} is nonempty, and it is closed under multiplication since if $gh_1 g^{-1}$ and $gh_2 g^{-1}$ are elements of

gHg^{-1}, then

$$(gh_1g^{-1})(gh_2g^{-1}) = gh_1h_2g^{-1},$$

which is in gHg^{-1} since $h_1h_2 \in H$. To check closure under inverses, note that $(ghg^{-1})^{-1} = gh^{-1}g^{-1}$, which is in gHg^{-1} since $h^{-1} \in H$. The verification that gHg^{-1} has the same number of elements as H is left as an exercise. $\square$

COROLLARY 11.5 If H is a subgroup of G, and no other subgroup has the same number of elements as H, then H is normal in G.

PROOF. For any $g \in G$, gHg^{-1} is a subgroup of G with the same number of elements as H, so gHg^{-1} must be H by hypothesis. Thus H is normal. $\square$

Examples

1. In S_3, $A_3 (= \langle f \rangle)$ is the only subgroup consisting of three elements. Hence we see again that A_3 is a normal subgroup of S_3.

2. For an example where neither of Theorems 11.2 or 11.3 is applicable but Corollary 11.5 is, let H be the subgroup

$$H = \{e, (1,2)(3,4), (1,3)(2,4), (1,4)(2,3)\}$$

in

$$G = A_4 = H \cup \{(1,2,3),(1,3,2),(1,2,4),(1,4,2),(1,3,4),(1,4,3),(2,3,4),(2,4,3)\}.$$

Since every element of $G - H$ has order 3, none of these elements can be contained in a subgroup of order 4. Thus H is the only subgroup of order 4, so H is normal in A_4, although H is not contained in $Z(A_4)$ and does not have index 2.

We will now proceed to the business that motivated the introduction of normal subgroups in the first place: we will show that if H is a normal subgroup of G, then there is a natural way to turn the set of right ($=$ left) cosets of H into a group.

First a standard piece of notation: We write $H \lhd G$ to signify that H is a normal subgroup of G.

Now let H be any subgroup of G. If we want to turn the set of right cosets of H into a group by using an operation that comes from the operation in G, there is really only one way to try to proceed. If we take two right cosets Ha and Hb and try to produce a right coset to be their product, what should it be? Hab, of course.

There is only one problem with this: namely, is it a well-defined operation? That is, suppose Ha is also Ha' and Hb is also Hb'. Is it then true that

Hab is the same thing as *Ha'b'*? In other words, is our intended operation independent of what representatives we pick from the cosets to perform the multiplication? It turns out that *it is if and only if H is a normal subgroup of G.*

Now saying that $Ha = Ha'$ is the same as saying that $a(a')^{-1} \in H$, by Corollary 9.4. Likewise, saying that $Hb = Hb'$ is just saying that $b(b')^{-1} \in H$. We want to see when these two assumptions imply that $Hab = Ha'b'$, that is, $(ab)(a'b')^{-1} \in H$. But

$$
\begin{aligned}
(ab)(a'b')^{-1} &= ab(b')^{-1}(a')^{-1} \\
&= a\big[b(b')^{-1}\big]a^{-1}a(a')^{-1} \\
&= \big(a\big[b(b')^{-1}\big]a^{-1}\big)a(a')^{-1}.
\end{aligned}
$$

Therefore, since $a(a')^{-1} \in H$, we see that $(ab)(a'b')^{-1} \in H$ if and only if $(a[b(b')^{-1}]a^{-1}) \in H$. But also $b(b')^{-1} \in H$; so certainly if H is normal then $a[b(b')^{-1}]a^{-1} \in H$ and we have what we want. On the other hand, having what we want for all possible choices of a and b *requires* that H be normal; that is, we need $a[b(b')^{-1}]a^{-1} \in H$ for all choices of a and b, and this forces H to be normal.

Thus we have established that the natural attempt at inducing an operation on the set of right cosets from that on G will succeed *if and only if H is a normal* subgroup. Therein lies the significance of the concept.

We now press on and show that if $H \lhd G$ then this operation on the set of right cosets turns the set into a group.

Notation. If $H \lhd G$, then G/H denotes the set of right ($=$ left) cosets of H in G.

THEOREM 11.6 If $H \lhd G$, then G/H is a group under the operation $Ha * Hb = H(ab)$.

PROOF. We have only to check that $*$ satisfies the associative law, that there is an identity element, and that inverses exist.

Observe that

$$
\begin{aligned}
(Ha * Hb) * Hc &= H(ab) * Hc = H[(ab)c] = H[a(bc)] \\
&= Ha * H(bc) = Ha * (Hb * Hc),
\end{aligned}
$$

so $*$ is in fact associative. Notice how $*$ is inheriting its associativity from the associativity of the operation in G.

The identity element of G/H is $He\ (=H)$, since for any $a \in G$ we have

$$
Ha * He = H(ae) = Ha,
$$

and similarly $He * Ha = Ha$.

Finally, the inverse of Ha is Ha^{-1} (what else?):

$$Ha * Ha^{-1} = H(aa^{-1}) = He,$$

the identity element, and similarly $Ha^{-1} * Ha = He$. □

The group G/H is called the **quotient group**, or **factor group**, of G by H. The symbol "G/H" is usually read "$G \bmod H$."

Observe that

$$|G/H| = [G:H].$$

Examples

1. Let G = Klein's 4-group = $\{e, a, b, c\}$ and let $H = \langle a \rangle$. Then H is normal since G is abelian, and G/H is a group with two elements in it, namely H and $Hb \, (= Hc)$.

2. Let $G = S_3$, and let $H = A_3 = \langle f \rangle$. Then H is normal in G and again G/H has two elements. They are H and $Hg \, (= Hfg = Hf^2 g)$. We have $Hg * Hg = Hg^2 = He = H$.

Let's look at how coset multiplication fails to be well defined if we take for H some *non-normal* subgroup of S_3. Try $H = \langle g \rangle = \{e, g\}$. The right cosets are H, $Hf = \{f, gf\}$, and $Hf^2 = \{f^2, gf^2\}$. Consider trying to multiply $Hf * Hf^2$, which is the same thing as $H(gf) * H(gf^2)$. On the one hand we get $Hf^3 = He$ $= H$; but using the other representatives we get

$$H(gfgf^2) = Hgffg = Hfgg = Hf,$$

not the same answer.

3. Let $G = Q_8$ and let $H = \{I, -I\}$. Then $H \lhd G$ and G/H is a group containing four elements, namely H, $H \cdot J$, $H \cdot K$, and $H \cdot L$. Therefore, as we remarked in Section 10, G/H must be essentially the same as either $(\mathbb{Z}_4, \oplus)$ or Klein's 4-group. Which one? Since

$$(H \cdot J)^2 = H \cdot J^2 = H \cdot -I = H,$$

and similarly, $(H \cdot K)^2 = H$ and $(H \cdot L)^2 = H$, the answer is Klein's 4-group, because every nonidentity element has order 2.

4. Let $G = (\mathbb{Z}, +)$ and let $H = \langle 2 \rangle$. Then G/H has two elements. More generally, let $G = (\mathbb{Z}, +)$ and let $H = \langle n \rangle$, where n is a positive integer. Then G/H has n elements, namely $H + 0, H + 1, H + 2, \ldots, H + (n-1)$. Observe that addition of these elements corresponds to addition of the indicated representatives modulo n. For instance, if $n = 6$, then $(H + 2) + (H + 5) = H + 7 =$ $H + 1$, corresponding to $2 \oplus 5 = 1$ in $(\mathbb{Z}_6, \oplus)$.

5. Let $G = \mathbb{Z}_3 \times \mathbb{Z}_2$, and let H be the normal subgroup $\{0\} \times \mathbb{Z}_2 = \{(0,0), (0,1)\}$. The elements of G/H are H, $H+(1,0)$, and $H+(2,0)$, and, for instance, we have

$$[H+(1,0)] + [H+(2,0)] = H+(0,0) = H.$$

In closing, we remark that quotient groups are very valuable tools. For example, in proving results about finite groups, one often proceeds by induction on the order of the group G. The crucial step is frequently to consider the quotient of G by some nontrivial normal subgroup. This quotient group is a group of smaller order than G, and hence is subject to the inductive hypothesis.

We shall use this approach repeatedly in Sections 14 and 15. For now, we offer the following illustration, which provides a partial converse to Lagrange's Theorem.

THEOREM 11.7 Let G be a finite abelian group and suppose p is a prime such that p divides $|G|$. Then G has a subgroup of order p.

PROOF (by induction on $|G|$). If $|G| = 1$, the result is trivial, since then no prime p divides $|G|$. Assume now that $|G| = m$, and that the result is true for all abelian groups of order less than m. Let $x \neq e$ be an element of G, and let $H = \langle x \rangle$. If $H = G$, then G is cyclic, and we are done by Theorem 5.5. Otherwise, H is a proper nontrivial normal subgroup of G (why?), and by Lagrange's Theorem we have

$$|G| = |G/H| \cdot |H|,$$

where both factors on the right are smaller than $|G|$. Now since p divides $|G|$ and p is prime, p divides either $|H|$ or $|G/H|$. If p divides $|H|$, then since H is abelian, the inductive hypothesis implies that H has a subgroup of order p, and this is the desired subgroup of G. Otherwise, p divides $|G/H|$, and since G/H is abelian (see Exercise 11.15), the inductive hypothesis implies that G/H has a subgroup of order p. This subgroup must be cyclic, so it is $\langle Hg \rangle$ for some element Hg of order p. By Exercise 11.9, $o(g) = kp$ for some integer k, and thus $o(g^k) = p$ by Theorem 4.4. Thus $\langle g^k \rangle$ is the subgroup we seek in G. $\square$

We know that Theorem 11.7 is no longer true if we drop the assumptions that G is abelian and p is prime. One outcome of Sections 14 and 15 will be that it is still true if we drop either one of these assumptions and keep the other in force.

EXERCISES

11.1 Show that $SL(2, \mathbb{R}) \lhd GL(2, \mathbb{R})$. [Recall that

$$SL(2, \mathbb{R}) = \left\{ \begin{pmatrix} a & b \\ c & d \end{pmatrix} \in GL(2, \mathbb{R}) | ad - bc = 1 \right\}.]$$

11.2 Let H be the subgroup of $GL(2, \mathbb{R})$ consisting of all matrices $\begin{pmatrix} a & b \\ 0 & d \end{pmatrix}$ such that $ad \neq 0$. Is H a normal subgroup of $GL(2, \mathbb{R})$?

11.3 Let $H \lhd G$, and assume that $|H| = 2$. Show that $H \subseteq Z(G)$.

11.4 Let $H \lhd G$ and $K \lhd G$. Show that $H \cap K \lhd G$.

11.5 Let G be a group, H a subgroup of G, and $K \lhd G$. Show that $H \cap K \lhd H$.

11.6 Let $H \lhd G$ and $K \lhd G$, and assume that $H \cap K = \{e\}$. Show that if $x \in H$ and $y \in K$, then $xy = yx$.

11.7 Let N be a normal subgroup of G and let H be any subgroup of G. Let $NH = \{nh | n \in N, h \in H\}$. Show that NH is a subgroup of G.

11.8 Suppose that in Exercise 11.7 we add the assumption that H is normal in G. Show that we can then conclude that NH is normal.

11.9 Let G be a group, let $g \in G$, and let $H \lhd G$. Suppose that the element $Hg \in G/H$ has order n. Show that if $o(g) = m$ then n divides m.

11.10 How many normal subgroups does D_4 have?

11.11 a) Let $G = D_4$. Show that there exist subgroups H and K of G such that $K \lhd H$ and $H \lhd G$, but $K \not\lhd G$.

b) Prove the same result for $G = A_4$ instead of D_4. (*Suggestion:* Let $H = \{e, (1,2)(3,4), (1,3)(2,4), (1,4)(2,3)\}$.)

11.12 Suppose that $A \lhd G$ and $B \lhd H$. Show that $A \times B \lhd G \times H$.

11.13 Show that $D_4/Z(D_4)$ is a group of order 4. Which one?

11.14 Show that $(\mathbb{Q}, +)/(\mathbb{Z}, +)$ is an infinite group, every element of which has finite order.

11.15 Let G be abelian and let H be a subgroup of G. Show that G/H is abelian.

11.16 Let G be cyclic, and let H be a subgroup of G.
a) H is normal in G. Why?
b) Show that G/H is cyclic.

11.17 Give an example of a nonabelian group G such that $G/Z(G)$ is:
a) abelian;
b) nonabelian.

11.18 Let G be a finite group and let H be a normal subgroup such that $[G:H] = 20$ and $|H| = 7$. Suppose $x \in G$ and $x^7 = e$. Show that $x \in H$.

11.19 Let G be a group and let H be a subgroup of index 2. Show that for every $a \in G$, $a^2 \in H$.

11.20 Use Exercise 11.19 to verify the remark made in Section 10 to the effect that A_4 has no subgroup of order 6. (Show that there are more than six squares in A_4.)

11.21 Let $H \triangleleft G$. Here is another way of viewing the introduction of a multiplication on G/H. We multiply right cosets as subsets: $Ha * Hb = \{h_1 a h_2 b \mid h_1, h_2 \in H\}$. The only question is whether or not this subset is again a right coset of H in G. Show that it is, and that in fact it is Hab, so that this "multiplication as subsets" coincides with that introduced in the text.

11.22 Verify the assertion in the proof of Theorem 11.4 that gHg^{-1} has the same number of elements as H.

11.23 Let H be a subgroup of G. Define $N(H) = \{g \in G \mid gHg^{-1} = H\}$. $N(H)$ is called the **normalizer** of H.

a) Show that $N(H)$ is a subgroup of G. (This is a repeat of Exercise 5.26.)

b) Show that $H \triangleleft N(H)$.

c) Show that if K is a subgroup of G and $H \triangleleft K$, then $K \subseteq N(H)$.

11.24 Let G be a group and let N be a normal subgroup of G. Assume that N is cyclic. Show that every subgroup of N is normal in G.

11.25 Show that if $G/Z(G)$ is cyclic, then G is abelian.

11.26 Let G be a group. The **commutator subgroup** G' of G is the smallest subgroup of G containing all the commutators, i.e., all elements of the form $x^{-1}y^{-1}xy$. Thus G' is the intersection of all the subgroups that contain all the commutators.

a) Show that G' consists of all elements of G that can be written as products of commutators. (In general, it need not be true that every element of G' is a commutator.)

b) Show that if K is any subgroup of G such that $G' \subseteq K$, then $K \triangleleft G$. Thus, in particular, $G' \triangleleft G$.

c) Show that G/G' is abelian.

d) Show that if K is any normal subgroup of G such that G/K is abelian, then $G' \subseteq K$.

11.27 Let G be an abelian group such that every $x \neq e$ in G has order 2. Let H be an abelian group such that every element of H has odd order. Show that $Q_8 \times G \times H$ is Hamiltonian. (*Suggestion:* Show that for any two elements x and y in $Q_8 \times G \times H$, xyx^{-1} is a power of y. The result of Exercise 6.14 may be useful.)

Conversely, it can be shown that every Hamiltonian group must have the form $Q_8 \times G \times H$, for some G and H with the indicated properties.

HOMOMORPHISMS

In this section we are going to study "sensible" functions from one group to another. One reason for doing this is that it reveals something of the dynamic nature of group theory, something of the way in which different groups can interact. Another reason is that this interaction often tells us a great deal about one of the groups involved: One learns about a given group by bringing into play its relationships with other groups.

The first thing we should do is make clear what the word "sensible" means in the last paragraph. A function from one group to another is, after all, a function from one set to another; such a function is "sensible" or "reasonable" if it takes into account the fact that groups are *more* than just sets. Groups have operations on them, and the reasonable functions to look at are those that behave themselves with respect to these operations.

By this we mean nothing more than the following. Suppose G and H are groups and $\varphi: G \rightarrow H$ is a function. Suppose $a, b \in G$. Then we have $\varphi(a) \in H$, $\varphi(b) \in H$, and $\varphi(ab) \in H$. What we want is that $\varphi(a)\varphi(b)$ should be the same as $\varphi(ab)$. That is, we want φ to respect the relationship among the elements a, b, and ab. We want multiplying a by b and then applying φ to yield the same element of H as first applying φ to each of a and b and then multiplying the answers.

Functions that "preserve structure" in this way are called *homomorphisms*.

DEFINITION Let G, H be groups and let $\varphi: G \rightarrow H$ be a function. Then φ is called a **homomorphism** if for every a and b in G we have

$$\varphi(ab) = \varphi(a)\varphi(b).$$

We emphasize that on the left-hand side of this equation the multiplication is performed in G; on the right-hand side, it is performed in H. And we repeat that the idea is that multiplying in G and then mapping over to H

yields the same result as first mapping to H and then multiplying.

Before giving some examples, we single out an especially important kind of homomorphism.

DEFINITIONS Let $\varphi : G \to H$ be a homomorphism. Then φ is called an **isomorphism** if it is a one-to-one onto function. Two groups G and H are said to be **isomorphic** if there exists an isomorphism from G onto H. If G and H are isomorphic, we write $G \cong H$.

It is worthwhile to point out a common misunderstanding of the definition of isomorphic groups. If we have tried several maps from G to H and found that none of them is an isomorphism, it does *not* follow that G and H are not isomorphic. For we might still find some other function that *is* an isomorphism. If there is any function from G to H that is an isomorphism, then $G \cong H$; to show that G and H are *not* isomorphic, we have to show that *no* function from G to H is an isomorphism. Theorems 12.4 and 12.5 will provide us with some ways of doing this.

Sometimes, one-to-one homomorphisms are called **monomorphisms**, and onto homomorphisms are called **epimorphisms**. In this terminology, a mapping is an isomorphism iff it is both a monomorphism and an epimorphism.

Examples

1. Let G be any group, and let $\varphi : G \to G$ be the identity map, $\varphi(x) = x$ for all $x \in G$. Then φ is an isomorphism.

Let $\varphi : G \to G$ be given by $\varphi(x) = e$ for all $x \in G$. Then φ is a homomorphism, but φ is neither a monomorphism nor an epimorphism, unless G is the trivial group.

2. Let $G = (\mathbb{Z}, +)$ and let $\varphi : G \to G$ be given by $\varphi(n) = 2n$ for all $n \in \mathbb{Z}$. Then φ is a homomorphism since

$$\varphi(n + m) = \varphi(n) + \varphi(m)$$

for all $n, m \in \mathbb{Z}$, that is,

$$2(n + m) = 2n + 2m.$$

φ is a monomorphism since if $2n = 2m$, then $n = m$. φ is not an epimorphism.

3. Again let $G = (\mathbb{Z}, +)$, and this time let φ be given by $\varphi(n) = -n$. φ is an isomorphism from G onto itself.

In general, an isomorphism from a group onto itself is called an **automorphism** ("auto" for "self"). The mapping $\varphi(n) = -n$ is called a **nontrivial automorphism**, because it is not the identity mapping. It can be shown that any group with more than two elements has a nontrivial automorphism.

4. Let $\varphi:(\mathbb{Z}, +)\rightarrow(\mathbb{Z}_n, \oplus)$ be given by $\varphi(m)=\overline{m}$, the remainder of m (mod n). φ is a homomorphism because for any integers m_1 and m_2,

$$\overline{m_1 + m_2} = \overline{m_1} \oplus \overline{m_2}.$$

φ is not a monomorphism since, for example, $\varphi(0)=\varphi(n)$. φ is an epimorphism, however.

5. Let $\varphi:(\mathbb{R}, +)\rightarrow(\mathbb{R}^+, \cdot)$ be given by $\varphi(x)=e^x$. φ is a homomorphism since

$$\varphi(x+y)=\varphi(x)\varphi(y),$$

that is,

$$e^{x+y}=e^x e^y.$$

In fact φ is both a monomorphism and an epimorphism, so $(\mathbb{R}, +)$ and $(\mathbb{R}^+, \cdot)$ are isomorphic.

6. Let G be Klein's 4-group. Define $\varphi:G\rightarrow G$ by $\varphi(e)=\varphi(a)=e$, $\varphi(b)=\varphi(c)=a$. φ is a homomorphism [for example, $\varphi(ab)=\varphi(c)=a=ea=\varphi(a)\varphi(b)$], but φ is neither one-to-one nor onto.

Now define φ by $\varphi(e)=\varphi(a)=\varphi(b)=e, \varphi(c)=a$. This φ is not a homomorphism, since $\varphi(ab)=\varphi(c)=a$, but $\varphi(a)\varphi(b)=ee=e\neq a$.

7. Define $\varphi: GL(2,\mathbb{R})\rightarrow(\mathbb{R}-\{0\}, \cdot)$ by

$$\varphi\left[\begin{pmatrix} a & b \\ c & d \end{pmatrix}\right]=\text{determinant of }\begin{pmatrix} a & b \\ c & d \end{pmatrix}=ad-bc.$$

Then φ is a homomorphism since the determinant of the product of two matrices is the product of their respective determinants, that is,

$$\varphi\left[\begin{pmatrix} a & b \\ c & d \end{pmatrix}\begin{pmatrix} e & f \\ g & h \end{pmatrix}\right]=\varphi\left[\begin{pmatrix} a & b \\ c & d \end{pmatrix}\right]\varphi\left[\begin{pmatrix} e & f \\ g & h \end{pmatrix}\right].$$

φ is an epimorphism, but not a monomorphism.

8. Let $\varphi:S_n\rightarrow(\mathbb{R}-\{0\}, \cdot)$ be given by

$$\varphi(f)=\begin{cases} 1 & \text{if } f \text{ is even} \\ -1 & \text{if } f \text{ is odd.} \end{cases}$$

Then φ is a homomorphism. For example, if f, g are both odd, then $\varphi(f)=\varphi(g)=-1$, and since fg is even, $\varphi(fg)=1$. Thus

$$\varphi(fg)=\varphi(f)\varphi(g).$$

The cases where f and g are both even, or one is even and one is odd, are handled similarly.

9. Let G and H be groups, and let $\varphi: G \times H \to H$ be given by $\varphi[(g,h)] = h$. φ is a homomorphism, since

$$\varphi[(g_1,h_1)(g_2,h_2)] = \varphi[(g_1 g_2, h_1 h_2)] = h_1 h_2 = \varphi[(g_1,h_1)]\varphi[(g_2,h_2)].$$

φ is called "projection onto the second component." Similarly, we have a projection homomorphism onto the first component.

It will be useful to know what happens when we apply two homomorphisms in succession.

THEOREM 12.1

i) Let $\varphi: G \to H$ and $\psi: H \to K$ be homomorphisms. Then $\psi \circ \varphi: G \to K$ is a homomorphism.

ii) If φ and ψ are both isomorphisms, so is $\psi \circ \varphi$.

iii) If $\varphi: G \to H$ is an isomorphism, so is $\varphi^{-1}: H \to G$.

PROOF. i) We must verify that for any $a, b \in G$,

$$(\psi \circ \varphi)(ab) = [(\psi \circ \varphi)(a)][(\psi \circ \varphi)(b)].$$

But

$$(\psi \circ \varphi)(ab) = \psi[\varphi(ab)] = \psi[\varphi(a)\varphi(b)]$$

$$= \psi[\varphi(a)]\psi[\varphi(b)] = [(\psi \circ \varphi)(a)][(\psi \circ \varphi)(b)].$$

ii) By part (i), $\psi \circ \varphi$ is a homomorphism. Since $\psi \circ \varphi$ is one-to-one and onto if both φ and ψ are, $\psi \circ \varphi$ is an isomorphism if both φ and ψ are.

iii) If φ is one-to-one and onto, then we know that φ^{-1} is a one-to-one onto function from H to G. We must verify that

$$\varphi^{-1}(xy) = \varphi^{-1}(x)\varphi^{-1}(y)$$

for all $x, y \in H$. Both sides of this equation represent elements of G, so since φ is one-to-one it suffices to show that

$$\varphi[\varphi^{-1}(xy)] = \varphi[\varphi^{-1}(x)\varphi^{-1}(y)].$$

But $\varphi[\varphi^{-1}(xy)] = xy$, and

$$\varphi[\varphi^{-1}(x)\varphi^{-1}(y)] = \varphi[\varphi^{-1}(x)]\varphi[\varphi^{-1}(y)] = xy \quad \text{too.} \quad \square$$

Example Let $\varphi: (\mathbb{R}, +) \to (\mathbb{R}^+, \cdot)$ be the isomorphism given by $\varphi(x) = e^x$. In this case, the inverse isomorphism $\varphi^{-1}: (\mathbb{R}^+, \cdot) \to (\mathbb{R}, +)$ is given by $\varphi^{-1}(x) = \ln x$. The fact that φ^{-1} is a homomorphism is expressed by the familiar equation

$$\ln xy = \ln x + \ln y.$$

One consequence of Theorem 12.1 is that the relation of isomorphism is an equivalence relation on the class of all groups. For instance, part (iii) assures us that it doesn't matter whether we say "$G \cong H$" or "$H \cong G$."

The concept of isomorphism provides us with a precise way of saying that two groups "look just alike" or "are essentially the same." For suppose $\varphi: G \rightarrow H$ is an isomorphism. If we replace each $a \in G$ by $\varphi(a)$, but keep the multiplication the same—that is, we define $\varphi(a)\varphi(b)$ to be $\varphi(c)$ iff $ab = c$ in G —then the group we get is precisely H. In some sense, we have just relabeled the elements of one group to get the other, and for all group-theoretic purposes the two groups are the same.

As an example, let $G = \{e_G, A, B, C\}$ be a noncyclic group of order 4. As we saw in Section 10, it must be the case that $A^2 = B^2 = C^2 = e_G$, $AB = BA = C$, $BC = CB = A$, and $AC = CA = B$, so that G "is essentially the same as" Klein's 4-group. The precise statement is that the mapping $\varphi: G \rightarrow V$ given by $\varphi(e_G) = e$, $\varphi(A) = a$, $\varphi(B) = b$, $\varphi(C) = c$ is an isomorphism. For instance, we have $\varphi(AB) = \varphi(C) = c = ab = \varphi(A)\varphi(B)$.

We also remarked in Section 10 that a group of order 4 which *is* cyclic looks just like $(\mathbb{Z}_4, \oplus)$, because it is just the set of powers of some element x of order 4. Saying this precisely, we have an isomorphism $\varphi: \mathbb{Z}_4 \rightarrow G$ given by $\varphi(n) = x^n$. More generally:

THEOREM 12.2 Let n be a positive integer, and let G be a cyclic group of order n. Then $G \cong (\mathbb{Z}_n, \oplus)$. Consequently, any two cyclic groups of order n are isomorphic to each other.

PROOF. Let $G = \langle g \rangle$, where $o(g) = n$. Then $G = \{e, g, g^2, \ldots, g^{n-1}\}$, all the indicated elements being distinct. Define $\varphi: \mathbb{Z}_n \rightarrow G$ by $\varphi(j) = g^j, 0 \leqslant j \leqslant n-1$. Then φ is an isomorphism from $\mathbb{Z}_n$ onto G. For clearly φ is one-to-one and onto; and φ is a homomorphism since for $j, k \in \mathbb{Z}_n$ we have

$$\varphi(j \oplus k) = g^{j \oplus k} = g^{j+k} = g^j g^k = \varphi(j)\varphi(k). \quad \square$$

The corresponding result for infinite cyclic groups is also true:

THEOREM 12.3 Let G be an infinite cyclic group. Then $G \cong (\mathbb{Z}, +)$. Consequently, any two infinite cyclic groups are isomorphic to each other.

PROOF. Exercise.

The next three theorems provide some information about the behavior of homomorphisms with respect to elements and subgroups.

THEOREM 12.4 Let $\varphi: G \to H$ be a homomorphism. Then

i) $\varphi(e_G) = e_H$;

ii) for any $x \in G$ and any integer n, $\varphi(x^n) = [\varphi(x)]^n$;

iii) if $o(x) = n$, then $o[\varphi(x)]$ divides n.

PROOF. i) We know that, in G, $e_G e_G = e_G$. Since φ is a homomorphism, this gives us $\varphi(e_G)\varphi(e_G) = \varphi(e_G)$, an equation in H. Since H is a group, this yields $\varphi(e_G) = e_H$, as desired.

ii) For $n = 0$ the result is just part (i), and for $n = 1$ it is trivial. We can finish the proof for positive n's by induction:

$$\varphi(x^n) = \varphi[x(x^{n-1})] = \varphi(x)\varphi(x^{n-1}) = \varphi(x)[\varphi(x)]^{n-1} = [\varphi(x)]^n.$$

Finally, if $n = -m$, with $m > 0$, then since $x^n x^m = e_G$, we get $\varphi(x^n)\varphi(x^m) = e_H$, so that $\varphi(x^n) = [\varphi(x^m)]^{-1}$ in H. By the case of positive exponents, $\varphi(x^m) = [\varphi(x)]^m$, so

$$\varphi(x^n) = \left[[\varphi(x)]^m \right]^{-1} = [\varphi(x)]^{-m} = [\varphi(x)]^n.$$

iii) If $x \in G$ and $o(x) = n$, then $x^n = e_G$, so by parts (i) and (ii), $[\varphi(x)]^n = e_H$. Thus $\varphi(x)$ has finite order, and $o[\varphi(x)]$ divides n by Theorem 4.4 (ii). $\square$

THEOREM 12.5 Let $\varphi: G \to H$ be an isomorphism. Then, in addition to the conclusions (i)–(iii) above, we have

iv) $o(x) = o[\varphi(x)]$, for every $x \in G$;

v) G and H have the same cardinality;

vi) G is abelian iff H is.

PROOF. iv) We have already observed in the proof of (iii) that for any n, $x^n = e_G$ implies $[\varphi(x)]^n = e_H$. In the present situation, the reverse is also true. For if $[\varphi(x)]^n = e_H$, that is, if $\varphi(x^n) = \varphi(e_G)$, then we must have $x^n = e_G$, because φ is now one-to-one.

Thus we have, for every n,

$$x^n = e_G \quad \text{iff} \quad [\varphi(x)]^n = e_H. \qquad [12.1]$$

It follows that there exists a positive n such that $x^n = e_G$ if and only if there exists a positive n such that $[\varphi(x)]^n = e_H$. Thus x has finite order iff $\varphi(x)$ does. And in the case where x and $\varphi(x)$ do have finite orders, Eq. [12.1] reveals that the *smallest* n such that $x^n = e_G$ equals the *smallest* n such that $[\varphi(x)]^n = e_H$. Thus $o(x) = o[\varphi(x)]$.

Of course, we have not used the fact that φ is onto in this proof, so statement (iv) remains valid if φ is merely a monomorphism.

v) This is obvious, since an isomorphism is a one-to-one onto function.

vi) Exercise. $\square$

Examples

1. Consider the homomorphism $\varphi:(\mathbb{Z}_8, \oplus) \to (\mathbb{Z}_4, \oplus)$ given by $\varphi(x) =$ remainder of $x \pmod 4$. Thus $\varphi(0) = \varphi(4) = 0$, $\varphi(1) = \varphi(5) = 1$, $\varphi(2) = \varphi(6) = 2$, and $\varphi(3) = \varphi(7) = 3$.

In $(\mathbb{Z}_8, \oplus)$, $o(1) = 8$. Now $\varphi(1) = 1 \in (\mathbb{Z}_4, \oplus)$, so $o[\varphi(1)] = 4$, which divides 8, in illustration of Theorem 12.4(iii). The fact that $o[\varphi(1)] \neq 8$ points out that φ is not an isomorphism, but of course this was obvious from the outset because φ as defined was not one-to-one. There can be no isomorphism from $(\mathbb{Z}_8, \oplus)$ onto $(\mathbb{Z}_4, \oplus)$, because these groups do not have the same number of elements.

2. $(\mathbb{Z}_4, \oplus)$ is not isomorphic to Klein's 4-group, since $(\mathbb{Z}_4, \oplus)$ contains an element of order 4, namely 1. If $\varphi:(\mathbb{Z}_4, \oplus) \to V$ were an isomorphism, then $\varphi(1)$ would be an element of order 4 in V. But there is no such element.

3. We have seen that $(\mathbb{R}, +)$ and $(\mathbb{R}^+, \cdot)$ are isomorphic. However, $(\mathbb{Q}, +)$ and $(\mathbb{Q}^+, \cdot)$ are not isomorphic. One reason is that every $q \in (\mathbb{Q}, +)$ has a "square root" $q/2$ in terms of addition, while some elements of $(\mathbb{Q}^+, \cdot)$ have no square root in terms of multiplication. For example, there is no $x \in \mathbb{Q}^+$ such that $x^2 = 3$. If $\varphi:(\mathbb{Q}, +) \to (\mathbb{Q}^+, \cdot)$ were an isomorphism, then there would be some $q \in \mathbb{Q}$ such that $\varphi(q) = 3$. We would have

$$[\varphi(q/2)]^2 = \varphi(q/2)\varphi(q/2) = \varphi(q/2 + q/2) = \varphi(q) = 3,$$

and $\varphi(q/2)$ would be a square root of 3 in $(\mathbb{Q}^+, \cdot)$. Since this is impossible, no such φ can exist.

4. $(\mathbb{R} - \{0\}, \cdot)$ and $GL(2, \mathbb{R})$ are not isomorphic, because one is abelian and the other is not.

THEOREM 12.6 Let $\varphi: G \to K$ be a homomorphism. Then:

i) If H is a subgroup of G, then $\varphi(H)$ is a subgroup of K. Here $\varphi(H)$ denotes the image of H, that is,

$$\varphi(H) = \{ k \in K \mid k \text{ is } \varphi(h) \text{ for some } h \in H \}.$$

ii) If J is a subgroup of K, then $\varphi^{-1}(J)$ is a subgroup of G. Here $\varphi^{-1}(J)$ is the **inverse image** of J, that is,

$$\varphi^{-1}(J) = \{ g \in G \mid \varphi(g) \in J \}.$$

iii) If $J \lhd K$, then $\varphi^{-1}(J) \lhd G$.

iv) Assume that φ is *onto*. Then if $H \lhd G$, $\varphi(H) \lhd K$.

A word of caution about the notation "$\varphi^{-1}(J)$": The use of the symbol "φ^{-1}" in this context does not imply that φ is one-to-one and has an inverse.

$\varphi^{-1}(J)$ makes sense here even if φ is wildly non-one-to-one. (The symbol "φ^{-1}" standing by itself, without "(J)" after it, only makes sense if φ has an inverse.)

PROOF OF THE THEOREM. i) We show that $\varphi(H)$ is a nonempty subset of K closed under multiplication and inverses. First, $\varphi(H)$ is nonempty since H is nonempty. Second, if $x,y \in \varphi(H)$, then $x = \varphi(a)$ and $y = \varphi(b)$ for some $a,b \in H$, so $xy = \varphi(a)\varphi(b) = \varphi(ab)$, and $\varphi(ab) \in \varphi(H)$ since $ab \in H$. Thus $\varphi(H)$ is closed under multiplication. Finally, if $x \in \varphi(H)$, then $x = \varphi(a)$ for some $a \in H$, so $x^{-1} = [\varphi(a)]^{-1} = \varphi(a^{-1}) \in \varphi(H)$, since $a^{-1} \in H$. Thus $\varphi(H)$ is closed under inverses.

   ii) Exercise.
   iii) Exercise.
   iv) We know from part (i) that $\varphi(H)$ is a subgroup of K. To show that $\varphi(H) \lhd K$, we must show that if $x \in K$ and $y \in \varphi(H)$, then $xyx^{-1} \in \varphi(H)$. Now $y = \varphi(a)$ for some $a \in H$, and since we are assuming that φ is *onto*, $x = \varphi(b)$ for some $b \in G$. Thus

$$xyx^{-1} = \varphi(b)\varphi(a)[\varphi(b)]^{-1} = \varphi(b)\varphi(a)\varphi(b^{-1}) = \varphi(bab^{-1}).$$

Now $bab^{-1} \in H$ since $H \lhd G$; thus $xyx^{-1} \in \varphi(H)$, and we are done. $\square$

Example We show how part (iv) may fail if φ is not onto.
   Let $G = (\mathbb{Z}_2, \oplus)$ and $K = S_3 = \{e, f, f^2, g, fg, f^2g\}$. Let $\varphi: G \to K$ be given by $\varphi(0) = e$, $\varphi(1) = g$. Clearly φ is a homomorphism, in fact a monomorphism. Now let $H = G$. Then $H \lhd G$, but $\varphi(H) = \langle g \rangle$, which is not a normal subgroup of K.

   We would like to use Theorem 12.6 to help us deliver on a promise we made in Section 8. In discussing symmetric groups, we remarked that one reason why they are important is that every group can be "thought of" as a subgroup of some symmetric group. The precise statement is that every group is isomorphic to a subgroup of some symmetric group.

THEOREM 12.7 (Cayley's Theorem) If G is a group, then G is isomorphic to a subgroup of S_G, the symmetric group on the set G.

PROOF. To define a mapping $\varphi: G \to S_G$, we must assign a permutation of G to each $g \in G$. Given g, define $f_g \in S_G$ by

$$f_g(x) = gx,$$

for all $x \in G$. We saw in Section 7 that f_g is a one-to-one mapping of G onto itself.

Now define $\varphi: G \to S_G$ by

$$\varphi(g) = f_g.$$

We assert that φ is a homomorphism, that is, $\varphi(g_1 g_2) = \varphi(g_1)\varphi(g_2)$, for all $g_1, g_2 \in G$. This equation says that

$$f_{g_1 g_2} = f_{g_1} \circ f_{g_2},$$

in other words, that $f_{g_1 g_2}$ and $f_{g_1} \circ f_{g_2}$ are the same element of S_G. To verify this, we show that $f_{g_1 g_2}(x) = (f_{g_1} \circ f_{g_2})(x)$, for every $x \in G$:

$$f_{g_1 g_2}(x) = (g_1 g_2)x = g_1(g_2 x) = f_{g_1}(g_2 x) = f_{g_1}(f_{g_2}(x)) = (f_{g_1} \circ f_{g_2})(x).$$

Since φ is a homomorphism, Theorem 12.6 tells us that $\varphi(G)$ is a subgroup of S_G (consisting of all the f_g's). G is isomorphic to this subgroup, since φ is one-to-one: if $\varphi(g_1) = \varphi(g_2)$, that is, if $f_{g_1} = f_{g_2}$, then in particular

$$f_{g_1}(e) = f_{g_2}(e)$$
$$g_1 e = g_2 e$$
$$g_1 = g_2. \quad \square$$

The impact of Cayley's Theorem is lessened somewhat by the fact that S_G is usually huge in relation to G. (For instance, if $|G| = 10$, then $|S_G| = 10! = 3,628,800$.) This makes it difficult to derive information about G from the fact that G is isomorphic to a subgroup of S_G. In Exercise 13.27 we will develop a generalization of Cayley's Theorem which sometimes enables us to show that G is isomorphic to a subgroup of a symmetric group smaller than S_G.

EXERCISES

12.1 Which of the following mappings are homomorphisms? Monomorphisms? Epimorphisms? Isomorphisms?

a) $G = (\mathbb{R} - \{0\}, \cdot)$, $H = (\mathbb{R}^+, \cdot)$; $\varphi: G \to H$ is given by $\varphi(x) = |x|$.

b) $G = (\mathbb{R}^+, \cdot)$; $\varphi: G \to G$ is given by $\varphi(x) = \sqrt{x}$.

c) $G = $ group of polynomials $p(x)$ with real coefficients, under addition of polynomials; $\varphi: G \to (\mathbb{R}, +)$ is given by $\varphi[p(x)] = p(1)$.

d) G is as in (c); $\varphi: G \to G$ is given by $\varphi[p(x)] = p'(x)$, the derivative of $p(x)$.

e) $G = $ the group of subsets of $\{1,2,3,4,5\}$ under symmetric difference; $A = \{1,3,4\}$, and $\varphi: G \to G$ is given by $\varphi(B) = A \triangle B$, for every $B \subseteq \{1,2,3,4,5\}$.

12.2 Let G be an abelian group. Show that the mapping $\varphi: G \to G$ given by $\varphi(x) = x^{-1}$ is an automorphism of G. Show that if G were not abelian, then φ would not be an automorphism.

12.3 Let G be an abelian group, let n be a positive integer, and let $\varphi: G \to G$ be given by $\varphi(x) = x^n$. Show that φ is a homomorphism. Need it be a monomorphism? An epimorphism?

12.4 In each case, determine whether or not the two given groups are isomorphic.

a) $(\mathbf{Z}_{12}, \oplus)$ and $(\mathbf{Q}^+, \cdot)$

b) $(2\mathbf{Z}, +)$ and $(3\mathbf{Z}, +)$

c) $(\mathbf{R} - \{0\}, \cdot)$ and $(\mathbf{R}, +)$

d) V and $\mathbf{Z}_2 \times \mathbf{Z}_2$

e) $\mathbf{Z}_3 \times \mathbf{Z}_3$ and $\mathbf{Z}_9$

f) $(\mathbf{R} - \{0\}, \cdot)$ and $(\mathbf{R}^+, \cdot) \times (\mathbf{Z}_2, \oplus)$

g) $(\mathbf{Z}, +)$ and $(\mathbf{Z}, *)$, where

$a * b = a + b - 1$

h) G and $G \times G$, where $G = \mathbf{Z}_2 \times \mathbf{Z}_2 \times \mathbf{Z}_2 \times \mathbf{Z}_2 \times \cdots$

(one copy of $\mathbf{Z}_2$ for each positive integer)

12.5 Let G and H be groups. Show that $G \times H \cong H \times G$.

12.6 Let G, H and K be groups. Show that $(G \times H) \times K \cong G \times H \times K$.

12.7 Show that if $A \cong G$ and $B \cong H$, then $A \times B \cong G \times H$.

12.8 Is $(\mathbf{Z}_{14}, \oplus)$ isomorphic to a subgroup of $(\mathbf{Z}_{35}, \oplus)$? Of $(\mathbf{Z}_{56}, \oplus)$?

12.9 Is V isomorphic to a subgroup of Q_8?

12.10 Let X be a set containing at least two elements. Show that V is isomorphic to a subgroup of $(P(X), \triangle)$.

12.11 Let $G = \mathbf{Z}_2 \times \mathbf{Z}_4$.

a) Find subgroups H and K of G such that $H \cong K$ but $G/H \not\cong G/K$.

b) Find subgroups A and B of G such that $G/A \cong G/B$ but $A \not\cong B$.

12.12 Show that there exist five groups of order 8, no two of which are isomorphic to each other.

12.13 Let $\varphi: G \to H$ be a homomorphism.

a) Show that if H is abelian and φ is one-to-one, then G is abelian.

b) Show that if G is abelian and φ is onto, then H is abelian.

c) Show that if φ is an isomorphism, then G is abelian iff H is.

12.14 Let $\varphi: G \to H$ be an isomorphism. Show that $Z(G) \cong Z(H)$.

12.15 Let $\varphi: G \to H$ be an onto homomorphism. Show that if G is cyclic, so is H.

12.16 Consider the mapping $\varphi: S_3 \to S_3$ given by $\varphi(f^i g^j) = f^{2i} g^j$. Show that φ is an automorphism of S_3.

12.17 How many automorphisms does Klein's 4-group have?

12.18 (Assumes familiarity with $n \times n$ matrices.) Let $GL(n, \mathbb{R})$ be the group of all invertible $n \times n$ real matrices under matrix multiplication. Let H be the subset of $GL(n, \mathbb{R})$ consisting of all matrices such that each column consists of one 1 and $(n-1)$ zeros, and each row consists of one 1 and $(n-1)$ zeros. Show that H is a subgroup of $GL(n, \mathbb{R})$ and $H \cong S_n$.

12.19 Prove Theorem 12.6(ii) and (iii).

12.20 (Exercise 12.3, revisited). Let G be a finite abelian group and let n be a positive integer relatively prime to $|G|$.

a) Show that the mapping $\varphi(x) = x^n$ is an automorphism of G.

b) Show that every $x \in G$ has an nth root, i.e., for every x there exists some $y \in G$ such that $y^n = x$.

12.21 Let G be the group of nonzero complex numbers under multiplication and let H be the subgroup of $GL(2, \mathbb{R})$ consisting of all matrices of the form $\begin{pmatrix} a & b \\ -b & a \end{pmatrix}$, where not both a and b are 0. Show that $G \cong H$.

12.22 Let G be a group and let $g \in G$. Show that the mapping $\varphi: G \to G$ given by $\varphi(x) = gxg^{-1}$ is an automorphism of G. It is called an **inner** automorphism, because it comes from an element of G.

12.23 A subgroup H of a group G is **characteristic** if $\varphi(H) \subseteq H$ for every automorphism φ of G.

a) Show that every characteristic subgroup is normal.

b) Show that the converse of (a) is false.

12.24 Suppose that $H \lhd G$ and K is a characteristic subgroup of H. Prove that $K \lhd G$. (See Exercises 12.22 and 12.23.)

12.25 Show that the center of a group is a characteristic subgroup. (See Exercise 12.23.)

12.26 Show that the commutator subgroup of a group is a characteristic subgroup. (See Exercise 11.26.)

12.27 If G is a group, $\mathrm{Aut}(G)$ denotes the set of automorphisms of G. Show that $\mathrm{Aut}(G)$ is a subgroup of $(S_G, \circ)$.

12.28 Let $G = (\mathbb{Z}_3, \oplus)$. Show that $\mathrm{Aut}(G)$ is not a normal subgroup of S_G.

12.29 Let p be a prime. Show that a cyclic group of order p has exactly $p - 1$ distinct automorphisms.

12.30 Let G be an infinite cyclic group. Prove that $\mathrm{Aut}(G) \cong (\mathbb{Z}_2, \oplus)$.

12.31 Let H be a proper subgroup of G and let ψ be an automorphism of H other than the identity mapping. Define a mapping $\varphi: G \to G$ by

$$\varphi(x) = \begin{cases} \psi(x) & \text{if } x \in H \\ x & \text{if } x \notin H. \end{cases}$$

Is φ an automorphism of G? Explain.

12.32 Let G and H be two isomorphic groups. Exhibit a one-to-one correspondence between the set of automorphisms of G and the set of isomorphisms from G onto H.

12.33 If we label the elements of $V = \{e, a, b, c\}$ with the integers $1, 2, 3, 4$, respectively, then the proof of Cayley's Theorem shows us how to find a subgroup of S_4 isomorphic to V. Write down the elements of this subgroup.

12.34 (See Exercise 12.22.) Show that the set of all inner automorphisms of a group G is a subgroup of $\text{Aut}(G)$. Is it a normal subgroup?

HOMOMORPHISMS AND NORMAL SUBGROUPS

In this section we will establish a connection between the seemingly unrelated concepts examined in the preceding two sections. The main idea is that all normal subgroups can be obtained from homomorphisms, and all homomorphisms can, in a sense, be obtained from normal subgroups.

First, let $H \triangleleft G$. We wish to find a homomorphism φ, defined on G, from which we can derive H. The first thing we need, in order to build φ, is an image group. That is, we need a group to replace the question mark:

$$\varphi: G \overset{\text{onto}}{\rightarrow} ?$$

This group had better involve H somehow, or else there will be little hope of recovering H from φ. We could try H itself, but easy examples show that there need not exist a homomorphism from G onto H (see Exercise 13.2). The only other thing that comes to mind is to take advantage of the fact that H is normal and try replacing the ? by G/H.

Is there a homomorphism from G onto G/H? There is certainly a natural mapping from G onto G/H, because we can send $a \in G$ to Ha. So let us define

$$\rho: G \rightarrow G/H$$

by

$$\rho(a) = Ha,$$

for every $a \in G$. Is ρ a homomorphism?

In other words, if $a, b \in G$ then is it true that $\rho(ab) = \rho(a)\rho(b)$, i.e., does

$$Hab = Ha * Hb?$$

Yes, it does, because this is the definition of the multiplication in G/H. The statement that ρ is a homomorphism may be taken as a formalization of our observation that the multiplication in G/H comes "naturally" from that in G.

ρ is called the **canonical** (or **natural**) homomorphism from G onto G/H.

Examples

1. Let $G = S_3 = \{e, f, f^2, g, fg, f^2g\}$, and let $H = \langle f \rangle$. Then G/H has order 2. If $\rho: G \to G/H$, then $\rho(e) = \rho(f) = \rho(f^2) = H = e_{G/H}$; and $\rho(g) = \rho(fg) = \rho(f^2g) = Hg$, because the elements g, fg, f^2g differ by elements of H. The canonical homomorphism sends everything in H to $e_{G/H}$, and thus "wipes out" differences that lie in H.

2. Let $G = Q_8$ and $H = \{I, -I\}$. If $\rho: G \to G/H$, then $\rho(I) = \rho(-I) = H = e_{G/H}$, and $\rho(J) = \rho(-J) = H \cdot J$. G/H has order 4, and as we saw in Section 11, every nonidentity element of G/H has order 2, so that G/H is isomorphic to Klein's 4-group. Thus there is a homomorphism from Q_8 onto V, and we say that V is a **homomorphic image** of Q_8.

Our next move is to try and recover H from ρ, and in fact this is very easy. If we are given ρ, then we get H by taking the set of elements in G that are mapped by ρ to the identity element in G/H. That is,

$$H = \{a \in G \mid \rho(a) = e_{G/H}\}.$$

The success of the preceding paragraph leads us to consider the general notion of the *kernel* of a homomorphism.

DEFINITION If $\varphi: G \to K$ is a homomorphism, then the **kernel** of φ is

$$\ker(\varphi) = \varphi^{-1}(\{e_K\}) = \{g \in G \mid \varphi(g) = e_K\}.$$

THEOREM 13.1 For any homomorphism $\varphi: G \to K$, $\ker(\varphi) \triangleleft G$.

PROOF. Since $\{e_K\} \triangleleft K$, this is a special case of Theorem 12.6, but for emphasis we repeat the proof that $\ker(\varphi)$ is normal. Let $g \in \ker(\varphi)$, $x \in G$. Then

$$\varphi(xgx^{-1}) = \varphi(x)\varphi(g)\varphi(x^{-1}) = \varphi(x)e_K[\varphi(x)]^{-1} = e_K,$$

so $xgx^{-1} \in \ker(\varphi)$, as desired. $\square$

Examples

1. Consider $\varphi: GL(2, \mathbb{R}) \to (\mathbb{R} - \{0\}, \cdot)$ given by $\varphi\left(\begin{pmatrix} a & b \\ c & d \end{pmatrix}\right) = ad - bc$. Then $\ker(\varphi) = \left\{\begin{pmatrix} a & b \\ c & d \end{pmatrix} \mid ad - bc = 1\right\} = SL(2, \mathbb{R})$, so $SL(2, \mathbb{R}) \triangleleft GL(2, \mathbb{R})$.

2. Consider $\varphi: S_n \to (\{1, -1\}, \cdot)$ given by

$$\varphi(f) = \begin{cases} 1 & \text{if } f \text{ is even} \\ -1 & \text{if } f \text{ is odd.} \end{cases}$$

Then $\ker(\varphi) = \{f \mid \varphi(f) = 1\} = A_n$. Thus we see again that $A_n \lhd S_n$.

We have seen that if we start with a normal subgroup H, form the canonical homomorphism ρ, and then take the kernel, we get back to H. The next theorem will show us what happens if we start with a homomorphism $\varphi: G \to K$, take its kernel, and then form the canonical homomorphism $\rho: G \to G/\ker(\varphi)$.

THEOREM 13.2 (Fundamental theorem on group homomorphisms) Let $\varphi: G \to K$ be a homomorphism from G *onto* K. Then $K \cong G/\ker(\varphi)$.

Remark. If $\varphi: G \to K$ is not necessarily onto, we get $\varphi(G) \cong G/\ker(\varphi)$.

PROOF OF THE THEOREM. We have a map φ from G to K and we wish to construct a map $\bar{\varphi}$ from $G/\ker(\varphi)$ to K. Let us write $\ker(\varphi) = N$ for simplicity. The elements of G/N are right cosets Na, and we have to decide where to send each such coset. What other try is there but $\bar{\varphi}(Na) = \varphi(a)$?

Let us reiterate. We have a mapping φ from G to K, and we wish to find one from G/N to K:

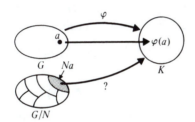

Our strategy in deciding where to send Na is to take a representative, a, for Na and see what φ does to it. $\varphi(a)$ will be an element of K, and this is where we send Na.

We have to check that this gives us a well-defined mapping, i.e., that if Na is also Nb then $\varphi(a) = \varphi(b)$, so that our definition of $\bar{\varphi}(Na)$ is independent of which representative of Na we pick to make the definition. Now if $Na = Nb$, then $ab^{-1} \in N = \ker(\varphi)$, so

$$\varphi(ab^{-1}) = e_K.$$

But

$$\varphi(ab^{-1}) = \varphi(a)\varphi(b^{-1}) = \varphi(a)[\varphi(b)]^{-1},$$

so $\varphi(a) = \varphi(b)$ and all is well.

$\bar{\varphi}$ is a homomorphism, since

$$\bar{\varphi}(NaNb) = \bar{\varphi}(Nab) = \varphi(ab) = \varphi(a)\varphi(b) = \bar{\varphi}(Na)\bar{\varphi}(Nb).$$

The crucial step in this chain of equalities is the fact that $\varphi(ab) = \varphi(a)\varphi(b)$. Thus the fact that $\bar{\varphi}$ is a homomorphism is thrown back on the fact that φ was one to begin with.

$\bar{\varphi}$ is one-to-one since if $\bar{\varphi}(Na) = \bar{\varphi}(Nb)$ then $\varphi(a) = \varphi(b)$, so $\varphi(ab^{-1}) = e_K$ and $ab^{-1} \in \ker \varphi = N$, yielding $Na = Nb$. Finally, $\bar{\varphi}$ is onto, since if $k \in K$ then there exists $a \in G$ such that $\varphi(a) = k$ (because φ was assumed to be onto), and this means that $\bar{\varphi}(Na) = k$. $\square$

Examples

1. Again let $\varphi: GL(2, \mathbb{R}) \rightarrow (\mathbb{R} - \{0\}, \cdot)$ be given by $\varphi\left(\begin{smallmatrix} a & b \\ c & d \end{smallmatrix}\right) = ad - bc$. Then φ is onto, since for any $r \neq 0$ we have $\left(\begin{smallmatrix} r & 0 \\ 0 & 1 \end{smallmatrix}\right) \in GL(2, \mathbb{R})$, and $\varphi\left(\left(\begin{smallmatrix} r & 0 \\ 0 & 1 \end{smallmatrix}\right)\right) = r$. By the Fundamental Theorem we conclude that

$$GL(2, \mathbb{R})/\ker(\varphi) \cong (\mathbb{R} - \{0\}, \cdot),$$

in other words,

$$GL(2, \mathbb{R})/SL(2, \mathbb{R}) \cong (\mathbb{R} - \{0\}, \cdot).$$

2. Let $\varphi: (\mathbb{Z}, +) \rightarrow (\mathbb{Z}_n, \oplus)$ be given by $\varphi(m) = \bar{m}$, the remainder of m (mod n). Then φ is onto, so the Fundamental Theorem says that

$$(\mathbb{Z}, +)/\ker(\varphi) \cong (\mathbb{Z}_n, \oplus),$$

that is,

$$(\mathbb{Z}, +)/n\mathbb{Z} \cong (\mathbb{Z}_n, \oplus).$$

This makes precise our observation in Section 11 that the addition of the cosets of $n\mathbb{Z}$ "corresponds" to the addition of their representatives, mod n.

3. Let $\varphi: S_n \rightarrow (\{1, -1\}, \cdot)$ be given by

$$\varphi(f) = \begin{cases} 1 & \text{if } f \text{ is even} \\ -1 & \text{if } f \text{ is odd}. \end{cases}$$

Then φ is onto, and as we have seen, $\ker(\varphi) = A_n$. Thus

$$S_n/A_n \cong (\{1, -1\}, \cdot).$$

4. Think of the complex numbers as the points in a plane by identifying the number $x + yi$ with the point (x, y). Let U be the set of points on the circle of radius 1 about the origin. Thus U consists of all points $x + yi$ such that $x^2 + y^2 = 1$, and the points in U are precisely those that can be represented in the form $\cos \theta + i \sin \theta$, for some real θ. We assert that U is a subgroup of $(\mathbb{C} - \{0\}, \cdot)$, and

$$U \cong (\mathbb{R}, +)/\mathbb{Z}.$$

To see this, define a mapping $\varphi: \mathbb{R} \to (\mathbb{C} - \{0\}, \cdot)$ by

$$\varphi(x) = \cos 2\pi x + i \sin 2\pi x.$$

φ maps $\mathbb{R}$ onto U, and φ is a homomorphism since for $x, y \in \mathbb{R}$ we have

$$\begin{aligned}
\varphi(x+y) &= \cos(2\pi x + 2\pi y) + i \sin(2\pi x + 2\pi y) \\
&= (\cos 2\pi x \cos 2\pi y - \sin 2\pi x \sin 2\pi y) \\
&\quad + i(\sin 2\pi x \cos 2\pi y + \cos 2\pi x \sin 2\pi y) \\
&= (\cos 2\pi x + i \sin 2\pi x)(\cos 2\pi y + i \sin 2\pi y) = \varphi(x)\varphi(y).
\end{aligned}$$

Thus U is a subgroup of $(\mathbb{C} - \{0\}, \cdot)$ by Theorem 12.6. Its identity element is $1 + 0i$, so

$$\ker(\varphi) = \{x \in \mathbb{R} | \cos 2\pi x = 1 \text{ and } \sin 2\pi x = 0\} = \mathbb{Z}.$$

By the Fundamental Theorem, $(\mathbb{R}, +)/\mathbb{Z} \cong U$.

5. Let G and H be groups, and consider the normal subgroup $G \times \{e_H\}$ in $G \times H$. It would seem that if we factor $G \times H$ by $G \times \{e_H\}$, we should get essentially H. Indeed we do, since there is an onto homomorphism $\varphi: G \times H \to H$ given by $\varphi[(g,h)] = h$, and the kernel of this map is $G \times \{e_H\}$. Thus

$$(G \times H)/(G \times \{e_H\}) \cong H.$$

These examples demonstrate that the Fundamental Theorem is a useful tool for obtaining isomorphisms. On the theoretical side, the theorem tells us that the image of any homomorphism can essentially be recovered from the kernel. Moreover, the proof of the theorem shows us how close we can come to recovering the homomorphism itself.

In the proof, the isomorphism $\bar{\varphi}$ was defined so that $\bar{\varphi}(Na) = \varphi(a)$ for every $a \in G$. Thus

$$\bar{\varphi}(\rho(a)) = \varphi(a)$$

for every a, or, in other words,

$$\bar{\varphi} \circ \rho = \varphi.$$

This equation is often expressed by saying that the diagram

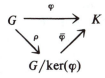

"commutes," because going directly from G to K accomplishes the same thing as taking the detour through $G/\ker(\varphi)$. The equation $\bar{\varphi} \circ \rho = \varphi$ tells us that φ and ρ come as close as could be hoped to being the "same" mapping. They map G onto groups that are isomorphic, and except for the isomorphism $\bar{\varphi}$ they are the same mapping.

In the situation of the Fundamental Theorem, there is a one-to-one correspondence between subgroups of K and subgroups of G that contain $\ker(\varphi)$, normal subgroups corresponding to normal subgroups. To discuss this correspondence, we shall use the fact that if H is a subgroup of G containing $\ker(\varphi)$ then $\varphi^{-1}[\varphi(H)] = H$. (For the notation, see the statement of Theorem 12.6.) To verify this fact, note that $H \subseteq \varphi^{-1}[\varphi(H)]$ automatically, that is, if $h \in H$ then φ maps h to an element of $\varphi(H)$. For the reverse inclusion, let $x \in \varphi^{-1}[\varphi(H)]$, i.e., suppose $\varphi(x) \in \varphi(H)$, with the aim of showing that $x \in H$. Since $\varphi(x) \in \varphi(H)$, we have $\varphi(x) = \varphi(h)$ for some $h \in H$. Thus $\varphi(xh^{-1}) = e_K$, so $xh^{-1} \in \ker(\varphi)$, and therefore $xh^{-1} \in H$ by our assumption that $\ker(\varphi) \subseteq H$. Thus $x = (xh^{-1})h$ is the product of two elements of H, so $x \in H$ as desired.

Now to any subgroup H of G containing $\ker(\varphi)$ we associate the subgroup $\varphi(H)$ of K; $\varphi(H)$ is a subgroup by Theorem 12.6. This association is one-to-one, since if H_1 is another subgroup of G containing $\ker(\varphi)$ and if $\varphi(H_1) = \varphi(H)$, then

$$\varphi^{-1}[\varphi(H_1)] = \varphi^{-1}[\varphi(H)],$$

hence $H_1 = H$ by the fact established above. The association is onto since if J is any subgroup of K, then $\varphi^{-1}(J)$ is a subgroup of G containing $\ker(\varphi)$, and $\varphi[\varphi^{-1}(J)]$ is J, because φ is onto (see Exercise 13.16). Notice that if $H \supseteq \ker(\varphi)$ is normal, then so is $\varphi(H)$; and if $\varphi(H)$ is normal in K, then $H \triangleleft G$ because $H = \varphi^{-1}[\varphi(H)]$. Here we are using parts (iii) and (iv) of Theorem 12.6.

We have proved

THEOREM 13.3 Let $\varphi: G \rightarrow K$ be an onto homomorphism. There is a one-to-one correspondence between the subgroups of K and those subgroups of G that contain $\ker(\varphi)$, given by $H \mapsto \varphi(H)$. We have $H \triangleleft G$ iff $\varphi(H) \triangleleft K$.

Observe that if H is *any* subgroup of G, not necessarily containing $\ker(\varphi)$, then $\varphi(H)$ is still a subgroup of K. But it is only by restricting ourselves to the subgroups of G containing $\ker(\varphi)$ that we get a one-to-one correspondence. For example, let $G = S_3 = \{e, f, f^2, g, fg, f^2g\}$, and let φ be the canonical homomorphism $G \rightarrow G/\langle f \rangle$. Since $|G/\langle f \rangle| = 2$, the only subgroups of $G/\langle f \rangle$ are the trivial subgroup and $G/\langle f \rangle$ itself. The corresponding subgroups of G containing $\langle f \rangle$ are $\langle f \rangle$ and G. However, there are proper subgroups H of G, not containing $\langle f \rangle$, such that $\varphi(H) = G/\langle f \rangle$ too. One example is $\langle g \rangle$; the others are $\langle fg \rangle$ and $\langle f^2g \rangle$.

The Fundamental Theorem is sometimes referred to as the *First* Isomorphism Theorem; we will now consider the *Second*. A bit of notation: If H and K are subgroups of G, then $HK = \{hk \mid h \in H, k \in K\}$.

THEOREM 13.4 (Second isomorphism theorem) Let H and K be subgroups of G, and assume $K \lhd G$. Then

$$H/(H \cap K) \cong HK/K.$$

The statement of this result is at first a bit forbidding, but it really makes good sense. If somebody asked you what HK/K looked like, you might say H; but after you thought a minute, you'd probably change your mind. When we factor K out of HK, some of H goes with it, namely $H \cap K$; so what is left is $H/(H \cap K)$.

PROOF. Let's first see that the terms in the conclusion make sense. To begin with, HK is a subgroup of G by Exercise 11.7. $K \lhd HK$ follows from $K \lhd G$. And $H \cap K \lhd H$ by Exercise 11.5.

Our strategy for the proof will be to use the Fundamental Theorem. We will define a homomorphism φ from H onto HK/K and show that its kernel is $H \cap K$.

For $h \in H$, define $\varphi(h) = Kh \in HK/K$. Observe that Kh is an element of HK/K, since $h \in HK$. φ is a homomorphism since

$$\varphi(h_1 h_2) = Kh_1 h_2 = Kh_1 Kh_2 = \varphi(h_1)\varphi(h_2).$$

φ is onto since any element of HK/K has the form $K(hk)$ for some $h \in H$, $k \in K$, and

$$K(hk) = Kh * Kk = Kh * K = Kh.$$

Thus $\varphi(h)$ is the given element $K(hk)$.

The Fundamental Theorem yields

$$H/\ker(\varphi) \cong HK/K,$$

so all that remains is to show that $\ker(\varphi) = H \cap K$. Now assume $x \in H$. Then

$$x \in \ker(\varphi) \Leftrightarrow \varphi(x) = e_{HK/K} \Leftrightarrow \varphi(x) = K \Leftrightarrow Kx = K \Leftrightarrow x \in K \Leftrightarrow x \in H \cap K.$$

This completes the proof. $\square$

Example Let $G = (\mathbb{Z}, +)$, $H = 4\mathbb{Z}$, $K = 6\mathbb{Z}$. The theorem says that

$$\frac{4\mathbb{Z}}{4\mathbb{Z} \cap 6\mathbb{Z}} \cong \frac{4\mathbb{Z} + 6\mathbb{Z}}{6\mathbb{Z}},$$

that is,

$$\frac{4\mathbb{Z}}{12\mathbb{Z}} \cong \frac{2\mathbb{Z}}{6\mathbb{Z}}.$$

This conclusion can be verified by other means, because both $4\mathbb{Z}/12\mathbb{Z}$ and $2\mathbb{Z}/6\mathbb{Z}$ are cyclic groups of order 3.

An application

Let G be a finite group, let H be a subgroup of G, and let $K \lhd G$. Theorem 13.4 tells us how many elements there are in HK. For it gives us

$$|H/(H \cap K)| = |HK/K|$$

$$\frac{|H|}{|H \cap K|} = \frac{|HK|}{|K|}$$

$$|HK| = \frac{|H||K|}{|H \cap K|}. \qquad [13.1]$$

One use for this information is that it often helps us investigate the structure of a given group. For instance, if we have a group G of order 15, a subgroup H of order 3, and a normal subgroup K of order 5, then $|H \cap K|$ divides both 3 and 5, hence is 1, so $|HK| = (3 \cdot 5)/1 = |G|$, and $G = HK$. We will use this kind of reasoning in Sections 14 and 15.

It is interesting to note that formula [13.1] for $|HK|$ remains valid without the assumption that K is normal, although in this case HK need not be a subgroup of G. We invite you to establish the general result in Exercise 13.14.

Theorem 13.4 has significant applications in the more advanced parts of group theory, but at the moment its main value for us is that it provides us with an opportunity to become more familiar with quotient groups and homomorphisms. The same remarks apply to our final result for this section, which points out one respect in which the quotient G/H really does behave like a fraction.

THEOREM 13.5 (Third isomorphism theorem) Suppose $H \lhd K \lhd G$ and $H \lhd G$. Then $K/H \lhd G/H$, and

$$\frac{G/H}{K/H} \cong G/K.$$

Remark. The assumption $H \lhd G$ is not redundant, because $H \lhd K \lhd G$ does not in general imply $H \lhd G$. See Exercise 11.11.

PROOF. First of all it is clear that $K/H \subseteq G/H$, that is, the set of right cosets $\{Ha | a \in K\}$ is contained in $\{Ha | a \in G\}$. In fact, K/H is obviously a subgroup of G/H, since K is a subgroup of G (Exercise 13.4). As for normality, let $Hk \in K/H$ and $Hg \in G/H$. Then, in G/H,

$$HgHk(Hg)^{-1} = HgHkHg^{-1} = Hgkg^{-1} \in K/H$$

since $gkg^{-1} \in K$ by the fact that $K \lhd G$. Thus $(G/H)/(K/H)$ makes sense.

To show that $(G/H)/(K/H) \cong G/K$, we shall again apply the Fundamental Theorem. This time we will define a homomorphism φ from G/H onto G/K and show that its kernel is K/H. Notice that this implies that K/H is normal all over again, but we didn't think it would hurt to check it at the outset.

To define φ, we must decide where in G/K to map Hg, for $g \in G$. Why not Kg? There is no other natural choice. But we do have to verify that φ, thus defined, is a well-defined mapping. That is, if $Hg = Hg_1$, is $Kg = Kg_1$, so that our definition is independent of our choice of a representative for Hg? The answer is an easy "yes," for if $Hg = Hg_1$, then $g(g_1)^{-1} \in H$, so $g(g_1)^{-1} \in K$, so $Kg = Kg_1$.

Note that φ is a homomorphism:

$$\varphi(Hg_1 Hg_2) = \varphi(Hg_1 g_2) = Kg_1 g_2 = Kg_1 Kg_2 = \varphi(Hg_1)\varphi(Hg_2).$$

And φ is onto since if $Kg \in G/K$, then we can find an element in G/H (namely, Hg) that gets mapped to Kg.

All that remains is to check that $\ker(\varphi) = K/H$. Now if $Hg \in G/H$, then

$$Hg \in \ker(\varphi) \Leftrightarrow \varphi(Hg) = e_{G/K} \Leftrightarrow \varphi(Hg) = K \Leftrightarrow Kg = K \Leftrightarrow g \in K \Leftrightarrow Hg \in K/H.$$

Thus $\ker(\varphi)$ is K/H, and since by the Fundamental Theorem

$$\frac{G/H}{\ker(\varphi)} \cong G/K,$$

we have the desired isomorphism. $\square$

Examples

  1. Let $G = (\mathbb{Z}, +)$, $K = 2\mathbb{Z}$, $H = 6\mathbb{Z}$. The theorem says that

$$\frac{(\mathbb{Z}, +)/6\mathbb{Z}}{2\mathbb{Z}/6\mathbb{Z}} \cong (\mathbb{Z}, +)/2\mathbb{Z}.$$

Observe that since $|2\mathbb{Z}/6\mathbb{Z}| = 3$, the index of $2\mathbb{Z}/6\mathbb{Z}$ in $(\mathbb{Z}, +)/6\mathbb{Z}$ (which has order 6) must be 2, so that the quotient group on the left-hand side has order 2. Since $(\mathbb{Z}, +)/2\mathbb{Z}$ also has order 2, what the theorem tells us in this case checks with our knowledge that any two groups of order 2 are isomorphic.

2. Let $G = Q_8 = \{I, -I, J, -J, K, -K, L, -L\}$, and consider the subgroups $\langle J \rangle = \{J, -I, -J, I\}$ and $\langle -I \rangle = \{-I, I\}$. The theorem says that

$$\frac{Q_8/\langle -I \rangle}{\langle J \rangle/\langle -I \rangle} \cong Q_8/\langle J \rangle.$$

Observe that the order of the quotient group on the left side is $4/2 = 2$ and the order of $Q_8/\langle J \rangle$ is $8/4 = 2$, so again we know by other means that these two groups are isomorphic.

EXERCISES

13.1 Let $\varphi: (\mathbb{Z}_8, \oplus) \rightarrow (\mathbb{Z}_4, \oplus)$ be given by $\varphi(x) =$ remainder of x (mod 4). Find $\ker(\varphi)$. To which familiar group is $(\mathbb{Z}_8, \oplus)/\ker(\varphi)$ isomorphic?

13.2 Let $G = Q_8$ and $H = \langle J \rangle \lhd G$. Show that there is no homomorphism from G onto H.

13.3 Find all the homomorphic images of D_4, up to isomorphism, by identifying each one with some familiar group.

13.4 Assume that $H \lhd K \lhd G$ and $H \lhd G$. Show that K/H is a subgroup of G/H.

13.5 Let G be the group of all real-valued functions on the real line, under addition of functions. Let H be the subset of G consisting of all f such that $f(0) = 0$.
a) Show that $H \lhd G$.
b) Show that $G/H \cong (\mathbb{R}, +)$.

13.6 Let

$$G = \frac{(\mathbb{Z}, +)/12\mathbb{Z}}{3\mathbb{Z}/12\mathbb{Z}}.$$

How many elements are there in G? Write them down explicitly.

13.7 Let $G = (\mathbb{C} - \{0\}, \cdot)$, and let U be the subgroup $U = \{x + yi \mid x^2 + y^2 = 1\}$. Use the Fundamental Theorem to show that:
a) $G/U \cong (\mathbb{R}^+, \cdot)$;
b) $G/\mathbb{R}^+ \cong U$.

13.8 Let $G = (\mathbb{Q} - \{0\}, \cdot)$, and let H be the subgroup $H = \{a/b \mid a$ and b are odd integers$\}$. Use the Fundamental Theorem to show that $G/H \cong (\mathbb{Z}, +)$.

13.9 Let G be a group, let $K \lhd G$, and let H be a subgroup of G such that $HK = G$ and $H \cap K = \{e\}$. Use the Second Isomorphism Theorem to show that $G/K \cong H$.

13.10 a) Let K be the subgroup $K = \{e, (1,2)(3,4), (1,3)(2,4), (1,4)(2,3)\}$ in S_4. Show that $K \lhd S_4$.
b) Show that $S_4/K \cong S_3$.

13.11 Let G and H be finite groups, and let $\varphi: G \to H$ be an onto homomorphism. Show that $|H|$ divides $|G|$.

13.12 Let m and n be positive integers. Show that there exists an onto homomorphism $\varphi: (\mathbb{Z}_n, \oplus) \to (\mathbb{Z}_m, \oplus)$ iff m divides n.

13.13 Let $A \lhd G$ and $B \lhd H$. Must it be true that

$$\frac{(G \times H)}{(A \times B)} \cong (G/A) \times (H/B)?$$

Either prove that it must, or give a counterexample.

13.14 Let H and K be subgroups of a finite group G. Prove that

$$|HK| = \frac{|H||K|}{|H \cap K|}.$$

13.15 Let H be a subgroup of G, and let $K \lhd G$. Show that $HK = KH$, and therefore the statement of the Second Isomorphism Theorem can also be written as

$$H/(H \cap K) \cong KH/K.$$

13.16 Let $\varphi: G \to K$ be an onto homomorphism and let J be a subgroup of K. Show that $\varphi[\varphi^{-1}(J)] = J$.

13.17 Let $\varphi: G \to K$ be a homomorphism. Prove that φ is one-to-one iff $\ker(\varphi) = \{e_G\}$.

13.18 Let $\varphi: G \to K$ be an epimorphism. Let $J \lhd K$. Prove that there exists a normal subgroup H of G such that $G/H \cong K/J$.

13.19 Let $\varphi: G \to K$ be an epimorphism, and let N be a subgroup of $\ker(\varphi)$ such that $N \lhd G$. Show that there is a homomorphism from G/N onto K.

13.20 Let $\varphi: G \to K$ be an epimorphism, and assume that K is abelian. Show that any subgroup of G containing $\ker(\varphi)$ is normal.

13.21 Prove Theorem 13.4 over again by using a φ that goes in the opposite direction from the one we used in the text. That is, define a mapping $\varphi: HK \to H/(H \cap K)$ by setting $\varphi(hk) = (H \cap K)h$. Show that φ is a well-defined homomorphism, and use the Fundamental Theorem to obtain the desired result.

13.22 Prove Theorem 13.5 over again by starting with a mapping from G onto $(G/H)/(K/H)$.

13.23 A group G is called **metabelian** if G has an abelian normal subgroup H such that G/H is also abelian.

a) Give an example of a metabelian group that is not abelian.

b) Suppose $\varphi: G \to K$ is an onto homomorphism and G is metabelian. Prove that K is metabelian.

13.24 (See Exercise 13.23.) Let G be a metabelian group. Show that every subgroup of G is metabelian.

13.25 Let G be an abelian group of order pq, where p and q are distinct primes. Show that G is cyclic.

13.26 Show that the group of inner automorphisms of a group G is isomorphic to $G/Z(G)$. (See Exercise 12.22 for the notion of "inner automorphism.")

13.27 (*Generalized Cayley's Theorem*). Let H be a subgroup of G and let X be the set of left cosets of H in G. Show that there exists a normal subgroup N of G such that $N \subseteq H$ and G/N is isomorphic to a subgroup of S_X. (*Suggestion*: Define a mapping $\varphi: G \to S_X$ by $\varphi(g) = f_g$, where $f_g(aH) = gaH$ for all $aH \in X$. Show that φ is a homomorphism and that $\ker(\varphi) \subseteq H$.)

13.28 Suppose $|G| = 10$ and G has a subgroup of order 2 which is not normal. Use the Generalized Cayley Theorem to show that G is isomorphic to a subgroup of S_5.

13.29 Let G be a group, H a subgroup of index n in G. Prove that there is a normal subgroup N of G such that $N \subseteq H$ and $[G:N]$ divides $n!$. (*Suggestion*: Use Exercise 13.27.)

13.30 (This exercise generalizes Theorem 11.3, for finite groups.) Suppose G is a finite group and p is the smallest prime that divides $|G|$. Let H be a subgroup of G such that $[G:H] = p$. Prove that $H \lhd G$. (*Suggestion*: Use Exercise 13.29.)

DIRECT PRODUCTS AND FINITE ABELIAN GROUPS

Up until now we have used direct products only as examples. But aside from providing us with an easy way of building new groups, they are also useful from another point of view in that they often enable us to understand a given group better. This happens when we are able to realize that the given group is isomorphic to the direct product of some of its subgroups. In this way, we can break the group down into simpler components that are easier to deal with.

Our goal in this section is to use direct products to analyze the structure of finite abelian groups. We will prove a classic result that gives us complete control over these groups.

To begin with, we seek to isolate conditions under which a group G will be isomorphic to the direct product of two of its subgroups, A and B. We can take a clue from $A \times B$ itself. For although A and B are not themselves subgroups of $A \times B$, there are subgroups A^* and B^* such that $A^* \cong A$, $B^* \cong B$, and thus $A \times B \cong A^* \times B^*$: take

$$A^* = A \times \{e_B\}, \quad B^* = \{e_A\} \times B.$$

What can we say about A^* and B^*?

For one thing, they are both normal in $A \times B$ (we checked this in Section 11). For another, $A^* B^* = A \times B$, because any element (a,b) in $A \times B$ can be written as $(a, e_B)(e_A, b)$, with $(a, e_B) \in A^*$ and $(e_A, b) \in B^*$. Finally, we have $A^* \cap B^* = \{(e_A, e_B)\} = \{e_{A \times B}\}$. It turns out that the three indicated properties capture the essence of the direct product situation.

THEOREM 14.1 Suppose A and B are subgroups of G such that

i) $A \triangleleft G$ and $B \triangleleft G$,

ii) $AB = G$,

iii) $A \cap B = \{e\}$.

Then $G \cong A \times B$.

PROOF. We first observe that (i)–(iii) imply two more properties of the subgroups A and B:

iv) If $ab = a_1 b_1$, where $a, a_1 \in A$ and $b, b_1 \in B$, then $a = a_1$ and $b = b_1$.

v) If $a \in A$ and $b \in B$, then $ab = ba$.

Observe that, whereas (ii) says that every element of G can be written as ab for some $a \in A$ and $b \in B$, (iv) says that this representation is unique. To prove (iv), note that $ab = a_1 b_1$ implies $a_1^{-1} a = b_1 b^{-1}$. Since $a_1^{-1} a \in A$, $b_1 b^{-1} \in B$, and $A \cap B = \{e\}$, we conclude that

$$a_1^{-1} a = e = b_1 b^{-1},$$

so $a = a_1$ and $b = b_1$. For (v), we want to show that $bab^{-1}a^{-1} = e$, so we show that $bab^{-1}a^{-1} \in A \cap B$ and apply (iii). Since $a \in A$ and $A \lhd G$, $bab^{-1}a^{-1} = (bab^{-1})a^{-1} \in A$; and since $b \in B$ and $B \lhd G$, $bab^{-1}a^{-1} = b(ab^{-1}a^{-1}) \in B$.

Now, to prove the theorem, define a function $\varphi: A \times B \to G$ by

$$\varphi[(a,b)] = ab.$$

φ is onto by (ii), and φ is one-to-one because if $\varphi[(a,b)] = \varphi[(a_1, b_1)]$, that is, if $ab = a_1 b_1$, then by (iv) $a = a_1$ and $b = b_1$, so $(a,b) = (a_1, b_1)$. Finally, φ is a homomorphism since

$$\varphi[(a,b)(a',b')] = \varphi[(aa', bb')] = aa'bb' = aba'b'$$

by (v), and this is $\varphi[(a,b)]\varphi[(a',b')]$. $\square$

Examples

1. Let $G = V = \{e, a, b, c\}$, and let $A = \{e, a\}$ and $B = \{e, b\}$. Then A and B are normal, $AB = \{ee, eb, ae, ab\} = \{e, b, a, c\} = V$, and $A \cap B = \{e\}$. Thus $V \cong A \times B$, and since A and B are both isomorphic to $\mathbb{Z}_2$, we have $V \cong \mathbb{Z}_2 \times \mathbb{Z}_2$. This checks with our knowledge that any noncyclic group of order 4 must be isomorphic to V.

2. Let $G = \langle x \rangle$ be a cyclic group of order mn, where $(m,n) = 1$. Then $A = \langle x^n \rangle$, $B = \langle x^m \rangle$ are normal subgroups of orders m and n, respectively. The order of any element of $A \cap B$ must divide both m and n, hence is 1, since $(m,n) = 1$. Thus $A \cap B = \{e\}$. Furthermore, $|AB| = (m \cdot n)/1 = mn$, so $AB = G$. Thus $G \cong A \times B$. This result agrees with our knowledge that $A \times B$ is cyclic (because $(m,n) = 1$), and any two cyclic groups of order mn are isomorphic to each other.

We now want to put Theorem 14.1 to work on finite abelian groups. A couple of definitions will help us state the result we are after.

If p is a prime number and G is a group, then G is said to be **of p-power order** if $|G| = p^k$ for some integer k. On the other hand, G is called a **p-group** if for every $x \in G$, $o(x)$ is a power of p. It follows from Theorems 10.4 and 11.7 that a finite abelian group is of p-power order iff it is a p-group. (We will see in the next section that the same is true for nonabelian groups.)

A group is **of prime-power order** if it is of p-power order for some prime p.

THEOREM 14.2 (Fundamental theorem on finite abelian groups) Let G be a nontrivial finite abelian group. Then G is isomorphic to the direct product of finitely many nontrivial cyclic groups of prime-power order. The prime-powers that occur as the orders of the factors are uniquely determined by G. More precisely, the primes that occur in the orders of the factors in any such decomposition of G are exactly the primes that divide $|G|$; and for any such prime p, if the orders of the factors that are p-groups in one such decomposition of G are $p^{t_1} \geqslant p^{t_2} \geqslant \cdots \geqslant p^{t_r}$, then the orders of the factors that are p-groups in *any* such decomposition of G are $p^{t_1} \geqslant p^{t_2} \geqslant \cdots \geqslant p^{t_r}$.

The uniquely determined integers $p^{t_1} \geqslant p^{t_2} \geqslant \cdots \geqslant p^{t_r}$, taken for all primes that divide $|G|$, are called the **invariants** of the nontrivial group G. We adopt the convention that the invariants of a trivial group are $\{1\}$.

Before we consider the proof of Theorem 14.2, we will list some corollaries which indicate the kind of control the theorem gives us over finite abelian groups. The first corollary tells us when two finite abelian groups are isomorphic.

COROLLARY 14.3 Let A and B be finite abelian groups. Then $A \cong B$ iff A and B have the same invariants.

PROOF. If A and B have the same invariants, then A is trivial iff B is. If they are both trivial, they are isomorphic. If neither is trivial, then when we write them both as direct products of (nontrivial) cyclic groups of prime-power order, we get the same number of factors of each order in both cases, and from this it follows that $A \cong B$.

Conversely, suppose $A \cong B$. If A is trivial, then so is B, so A and B have the same invariants. If A is not trivial and we write $A \cong A_1 \times A_2 \times \cdots \times A_r$, then $B \cong A_1 \times A_2 \times \cdots \times A_r$, so again A and B have the same invariants. $\square$

If n is a positive integer, then by a **partition** of n we mean a sequence of positive integers $t_1 \geqslant t_2 \geqslant \cdots \geqslant t_r$ such that $t_1 + t_2 + \cdots + t_r = n$. The number of distinct partitions of n is denoted by $p(n)$. For example, $p(4) = 5$, since we can write 4 as 4, $3+1$, $2+2$, $2+1+1$, or $1+1+1+1$.

COROLLARY 14.4 If q is a prime and n is a positive integer, then the number of nonisomorphic abelian groups of order q^n is $p(n)$. If $m = q_1^{n_1} q_2^{n_2} \cdots q_k^{n_k}$ for distinct primes q_i, then the number of nonisomorphic abelian groups of order m is $p(n_1)p(n_2)\cdots p(n_k)$.

PROOF OF THE FIRST STATEMENT. To any partition $t_1 \geqslant t_2 \geqslant \cdots \geqslant t_r$ of n, we associate the group

$$\mathbb{Z}_{q^{t_1}} \times \mathbb{Z}_{q^{t_2}} \times \cdots \times \mathbb{Z}_{q^{t_r}},$$

an abelian group of order q^n with invariants $q^{t_1} \geqslant q^{t_2} \geqslant \cdots \geqslant q^{t_r}$. The groups associated to different partitions are nonisomorphic because they have different invariants. Thus we get $p(n)$ nonisomorphic groups. But any abelian group of order q^n must be isomorphic to one of these because it has the same invariants as one of them. Hence there are precisely $p(n)$ nonisomorphic abelian groups of order q^n.

The proof of the second statement is left to the reader, but the idea should be clear from the following examples. $\square$

Examples

1. Let's find all abelian groups G of order 36. Since $36 = 2^2 \cdot 3^2$, the possibilities for the 2-groups in the prime-power decomposition of G are $\mathbb{Z}_{2^2}$ and $\mathbb{Z}_{2^1} \times \mathbb{Z}_{2^1}$, that is, $\mathbb{Z}_4$ and $\mathbb{Z}_2 \times \mathbb{Z}_2$. The possibilities for the 3-groups are $\mathbb{Z}_9$ and $\mathbb{Z}_3 \times \mathbb{Z}_3$. Thus we have

$$\mathbb{Z}_4 \times \mathbb{Z}_9,$$
$$\mathbb{Z}_4 \times \mathbb{Z}_3 \times \mathbb{Z}_3,$$
$$\mathbb{Z}_2 \times \mathbb{Z}_2 \times \mathbb{Z}_9,$$
$$\mathbb{Z}_2 \times \mathbb{Z}_2 \times \mathbb{Z}_3 \times \mathbb{Z}_3,$$

a total of four different abelian groups of order 36.

2. To find all abelian groups of order 600, we write $600 = 2^3 \cdot 3^1 \cdot 5^2$. The possibilities for the 2-groups are $\mathbb{Z}_8$, $\mathbb{Z}_4 \times \mathbb{Z}_2$, and $\mathbb{Z}_2 \times \mathbb{Z}_2 \times \mathbb{Z}_2$. There must be one 3-group, $\mathbb{Z}_3$. The 5-groups can be $\mathbb{Z}_{25}$ or $\mathbb{Z}_5 \times \mathbb{Z}_5$. So we have

$$\mathbb{Z}_8 \times \mathbb{Z}_3 \times \mathbb{Z}_{25},$$
$$\mathbb{Z}_8 \times \mathbb{Z}_3 \times \mathbb{Z}_5 \times \mathbb{Z}_5,$$
$$\mathbb{Z}_4 \times \mathbb{Z}_2 \times \mathbb{Z}_3 \times \mathbb{Z}_{25},$$
$$\mathbb{Z}_4 \times \mathbb{Z}_2 \times \mathbb{Z}_3 \times \mathbb{Z}_5 \times \mathbb{Z}_5,$$
$$\mathbb{Z}_2 \times \mathbb{Z}_2 \times \mathbb{Z}_2 \times \mathbb{Z}_3 \times \mathbb{Z}_{25},$$
$$\mathbb{Z}_2 \times \mathbb{Z}_2 \times \mathbb{Z}_2 \times \mathbb{Z}_3 \times \mathbb{Z}_5 \times \mathbb{Z}_5.$$

One of these groups must be isomorphic to $\mathbb{Z}_{600}$. In fact, it is $\mathbb{Z}_8 \times \mathbb{Z}_3 \times \mathbb{Z}_{25}$, because this is the only cyclic group in the list, by Theorem 6.1. We can also write $\mathbb{Z}_{600} \cong \mathbb{Z}_{24} \times \mathbb{Z}_{25} \cong \mathbb{Z}_8 \times \mathbb{Z}_{75} \cong \mathbb{Z}_3 \times \mathbb{Z}_{200}$. These examples serve to emphasize the fact that two products of nontrivial cyclic groups can be isomorphic without the number of factors being the same, and without the orders of the factors being the same. By restricting ourselves to factors of prime-power order, however, we achieve uniqueness.

The information provided by Theorem 14.2 can sometimes be used to handle the "abelian case" in more general contexts. For example, let us sketch a proof that there are essentially only five groups of order 8 (nonabelian ones included).

We assert that any group G of order 8 must be isomorphic to one of

$$\mathbb{Z}_8, \mathbb{Z}_4 \times \mathbb{Z}_2, \mathbb{Z}_2 \times \mathbb{Z}_2 \times \mathbb{Z}_2, Q_8, D_4.$$

If G is abelian, then we know that G is isomorphic to $\mathbb{Z}_8$, $\mathbb{Z}_4 \times \mathbb{Z}_2$, or $\mathbb{Z}_2 \times \mathbb{Z}_2 \times \mathbb{Z}_2$. If G is not abelian, then G can have no element of order 8 (else it would be cyclic), and G *must* have an element of order 4, since otherwise we would have $x^2 = e$ for every $x \in G$, making G abelian (Exercise 3.11). Say $a \in G$ and $o(a) = 4$. Then $\langle a \rangle$ is normal since it has index 2. Let $b \in G - \langle a \rangle$. Then $G = \{e, a, a^2, a^3, b, ab, a^2b, a^3b\}$. Now bab^{-1} must be a^3, since it has to be in $\langle a \rangle$ and cannot be e, a, or a^2 (Exercise 14.5). Thus $ba = a^3b$, and therefore the multiplication in G is completely determined once we know $o(b)$. But $o(b)$ must be either 2 or 4; if it is 2, then $G \cong D_4$, and if it is 4, then $G \cong Q_8$ (Exercise 14.5).

As our next corollary to Theorem 14.2, we will show that, for abelian groups, Lagrange's Theorem has a converse.

COROLLARY 14.5 Let G be an abelian group of order n and let m be an integer that divides n. Then G has a subgroup of order m.

PROOF. By Theorem 14.2, we can assume that

$$G = G_1 \times G_2 \times \cdots \times G_k,$$

where G_i is a cyclic group of order $p_i^{r_i}$, the p_i's being (not necessarily distinct) primes. Then $n = p_1^{r_1} p_2^{r_2} \cdots p_k^{r_k}$, so since $m | n$ we have $m = p_1^{s_1} p_2^{s_2} \cdots p_k^{s_k}$ for some integers s_i, $0 \leq s_i \leq r_i$. By Theorem 5.5 we can let H_i be a subgroup of G_i of order $p_i^{s_i}$; then

$$H_1 \times H_2 \times \cdots \times H_k$$

is a subgroup of G of order m. □

COROLLARY 14.6 Let G be a finite group with more than two elements. Then G has a nontrivial automorphism.

PROOF. Of course this is a statement about arbitrary finite groups, not just abelian ones. We will use Theorem 14.2 to handle the abelian case.

First suppose G is nonabelian, and choose $g \in G - Z(G)$. Then the mapping $\varphi: G \rightarrow G$ given by $\varphi(x) = gxg^{-1}$ is a nontrivial automorphism of G, because if we choose x such that $gx \neq xg$, then $\varphi(x) \neq x$.

If G is abelian, then the mapping $\varphi(x) = x^{-1}$ is an automorphism of G. It is nontrivial unless $x^{-1} = x$ for all $x \in G$, that is, unless every nonidentity element of G has order 2.

In this exceptional case, all the cyclic groups in the prime-power decomposition of G must have order 2, so we can write

$$G \cong \mathbb{Z}_2 \times \mathbb{Z}_2 \times \cdots \times \mathbb{Z}_2,$$

where there are at least two copies of $\mathbb{Z}_2$ because $|G| > 2$. The group $\mathbb{Z}_2 \times \mathbb{Z}_2 \times \cdots \times \mathbb{Z}_2$ has a nontrivial automorphism φ, obtained by interchanging the first and second components of all the elements. If ψ denotes an isomorphism from G onto $\mathbb{Z}_2 \times \mathbb{Z}_2 \times \cdots \times \mathbb{Z}_2$, then $\psi^{-1} \circ \varphi \circ \psi$ is a nontrivial automorphism of G. $\square$

The only place where we used the finiteness of G in this proof was in obtaining an isomorphism from G onto a product of $\mathbb{Z}_2$'s. It can be shown that if G is an infinite abelian group such that $x^2 = e$ for all $x \in G$, then G is isomorphic to a subgroup of the product of infinitely many $\mathbb{Z}_2$'s (the subgroup consisting of all the elements with only finitely many nonzero components). Our proof then shows that every group with more than two elements has a nontrivial automorphism.

The preceding corollaries demonstrate that Theorem 14.2 is a powerful tool. We will now take up its proof, which is essentially a reprise of the "quotient groups and induction" theme we introduced in Section 11.

The proof is somewhat longer than any we have done before, so we will split it into three steps:

Step 1. We show that every finite abelian group is isomorphic to a product of abelian p-groups.

Step 2. We show that every finite abelian p-group is isomorphic to a product of cyclic groups of p-power order.

Steps 1 and 2 establish the existence of the prime-power decomposition.

Step 3. We show that the prime-power decomposition is unique.

Suppose G is abelian and $|G| = p_1^{r_1} p_2^{r_2} \cdots p_k^{r_k}$, where the p_i's are distinct primes, and each $r_i \geqslant 1$. For each i, the set

$$G(p_i) = \left\{ x \in G \mid x^{p_i^{r_i}} = e \right\}$$

is a subgroup of G, and is a p_i-group. Step 1 of our program will be accomplished if we can show that

$$G \cong G(p_1) \times G(p_2) \times \cdots \times G(p_k),$$

and to do this it will suffice, by induction, to prove the following lemma.

LEMMA 14.7 Let G be an abelian group, and let $|G| = mn$, with $(m, n) = 1$. Let $A = \{x \in G \mid x^m = e\}$ and let $B = \{x \in G \mid x^n = e\}$. Then $G \cong A \times B$.

PROOF. A and B are subgroups of G, and they are both normal, because G is abelian. We must show that $AB = G$ and $A \cap B = \{e\}$.

Since $(m, n) = 1$, there are integers r and s such that $rn + sm = 1$. If $x \in G$, then

$$x = x^1 = x^{rn + sm} = x^{rn} x^{sm},$$

and $x^{rn} \in A$, $x^{sm} \in B$, since $mn = |G|$. Thus $G = AB$.

If $x \in A \cap B$, then $x = x^{rn} x^{sm} = ee = e$. $\square$

Step 2 of the proof takes a little more doing, but the plan of attack is straightforward. Suppose G is a finite abelian p-group; we wish to show by induction on $|G|$ that G is isomorphic to a product of cyclic groups of p-power order. If $|G| = 1$, then G is already cyclic of order p^0. Now assume the result is true for all p-groups of order less than $|G|$.

We want to choose $x \neq e$ in G, let $A = \langle x \rangle$, and find a subgroup B of G such that we can apply Theorem 14.1 to A and B. If we can do this, then since $|B| < |G|$, the inductive hypothesis will finish the proof.

Now if B exists at all, it has to be isomorphic to G/A, so it is natural to look at G/A. The inductive hypothesis also applies to this group, so we have

$$G/A \cong \langle y_1 \rangle \times \langle y_2 \rangle \times \cdots \times \langle y_m \rangle,$$

where $y_1, \ldots, y_m$ have orders $p^{t_1}, \ldots, p^{t_m}$, for some integers $t_1, \ldots, t_m$. This means that there are cosets $Ax_1, \ldots, Ax_m \in G/A$ such that:

Every element of G/A has a unique expression in the form

$$(Ax_1)^{r_1} (Ax_2)^{r_2} \cdots (Ax_m)^{r_m}, \quad \text{with } 0 \leqslant r_i < p^{t_i} \text{ for } 1 \leqslant i \leqslant m. \tag{14.1}$$

The way is now clear; we must try to show that we can choose representatives x_i for Ax_i such that $o(x_i) = o(Ax_i)$, and then

$$B = \left\{ x_1^{r_1} x_2^{r_2} \cdots x_m^{r_m} \mid 0 \leqslant r_i < p^{t_i} \right\}$$

will be our choice for B, isomorphic to G/A.

Assuming for the moment that the x_i's can be found, it is clear that B is a subgroup of G (B is a finite subset of G, closed under multiplication). We assert that $AB = G$ and $A \cap B = \{e\}$.

If $g \in G$, then by [14.1] we have

$$Ag = (Ax_1)^{r_1}(Ax_2)^{r_2} \cdots (Ax_m)^{r_m} = Ax_1^{r_1}x_2^{r_2} \cdots x_m^{r_m} = Ab,$$

with $b \in B$. Thus gb^{-1} is some $a \in A$, and $g = ab$, proving $G = AB$.

To see that $A \cap B = \{e\}$, suppose that $a \in A$ and $a = x_1^{r_1}x_2^{r_2} \cdots x_m^{r_m}$, with $0 \leqslant r_i < p^{t_i}$. Then in G/A we have

$$e_{G/A} = (Ax_1)^{r_1}(Ax_2)^{r_2} \cdots (Ax_m)^{r_m}.$$

Since we also have

$$e_{G/A} = (Ax_1)^0(Ax_2)^0 \cdots (Ax_m)^0,$$

the uniqueness in [14.1] implies that $r_1 = r_2 = \cdots = r_m = 0$. Thus $a = x_1^0 x_2^0 \cdots x_m^0 = e$, and we have what we want.

Everything has now come down to showing that we can choose representatives x_i such that $o(x_i) = o(Ax_i) = p^{t_i}$. To simplify the notation a bit, suppose we have a coset Ay with order p^t, and we want to find a representative with order p^t. Certainly p^t divides the order of *any* representative; the problem is to find a representative whose p^t-th power is e. Now since Ay has order p^t, we have $y^{p^t} \in A$, that is,

$$y^{p^t} = x^n, \qquad [14.2]$$

for some $n < o(x)$. If we had $n = cp^t$ for some c, we would get

$$y^{p^t} = (x^c)^{p^t},$$

and thus

$$(yx^{-c})^{p^t} = e.$$

Since yx^{-c} is a representative for Ay, we would be done.

Suppose p^w is the highest power of p that divides n; we want to show that $w \geqslant t$. We can get some information about the relationship between w and t by computing orders on both sides of Eq. [14.2]. Say $o(x) = p^i$, and $o(y) = p^j$; note that since p^t divides $o(y)$, $j \geqslant t$, and since $n < o(x)$, $w < i$. Now

$$o(y^{p^t}) = \frac{p^j}{(p^j, p^t)} = \frac{p^j}{p^t} = p^{j-t},$$

and

$$o(x^n) = \frac{p^i}{(p^i, n)} = \frac{p^i}{(p^i, p^w)} = \frac{p^i}{p^w} = p^{i-w}.$$

Thus $p^{j-t} = p^{i-w}$, so $j - t = i - w$, and

$$w = t + i - j.$$

To achieve our goal of showing that $w \geqslant t$, we need $i \geqslant j$, that is, we want $o(x) \geqslant o(y)$. As things stand this need not be true, but we can make it true by making a special choice of x at the outset. We choose x so that $o(x) \geqslant o(y)$ for all $y \in G$ to begin with, and then Step 2 of our proof is complete.

For Step 3, observe that if $G \cong G_1 \times \cdots \times G_k$, where $G_1, \ldots, G_k$ are nontrivial cyclic groups of prime-power order, then $|G| = |G_1| \cdot \cdots \cdot |G_k|$, so the primes that occur in the orders of the G_i's are precisely the primes that divide $|G|$. Also note that, for any such prime p, the product of those G_i's that are p-groups is isomorphic to the subgroup of G consisting of those elements with order a power of p. Thus, to establish the uniqueness of the prime-power decomposition, it will suffice to handle the case of a nontrivial finite p-group.

So suppose that

$$\langle x_1 \rangle \times \langle x_2 \rangle \times \cdots \times \langle x_r \rangle \cong \langle y_1 \rangle \times \langle y_2 \rangle \times \cdots \times \langle y_s \rangle, \qquad [14.3]$$

where $o(x_i) = p^{t_i}$, $t_1 \geqslant t_2 \geqslant \cdots \geqslant t_r \geqslant 1$, and $o(y_j) = p^{u_j}$, $u_1 \geqslant u_2 \geqslant \cdots \geqslant u_s \geqslant 1$. We proceed by induction on the (equal) orders of the two products involved, the case of order $p = p^1$ being trivial. For the induction step, notice that any isomorphism which gives us [14.3] must map the set of pth powers of all the elements on the left onto the set of pth powers of all the elements on the right. That is, [14.3] entails

$$\langle x_1^p \rangle \times \langle x_2^p \rangle \times \cdots \times \langle x_{i_0}^p \rangle \cong \langle y_1^p \rangle \times \langle y_2^p \rangle \times \cdots \times \langle y_{j_0}^p \rangle,$$

where i_0 is the largest i such that $t_i \geqslant 2$, and similarly for j_0. These new products have smaller order than the ones we started with, since $\langle x_i^p \rangle$ has order $p^{t_i - 1}$ and $\langle y_j^p \rangle$ has order $p^{u_j - 1}$. Thus the inductive hypothesis implies that $i_0 = j_0$ and

$$t_1 - 1 = u_1 - 1, \quad t_2 - 1 = u_2 - 1, \ldots, t_{i_0} - 1 = u_{j_0} - 1.$$

Therefore

$$t_1 = u_1, \ldots, t_{i_0} = u_{j_0},$$

and all that remains is to show that the number of i's with $t_i = 1$ equals the number of j's with $u_j = 1$, that is, $r - i_0 = s - j_0$. But if we compute orders on both sides of [14.3], we get

$$p^{t_1} p^{t_2} \cdots p^{t_{i_0}} p^{r - i_0} = p^{t_1} p^{t_2} \cdots p^{t_{i_0}} p^{s - j_0},$$

so $p^{r - i_0} = p^{s - j_0}$, and we are done. $\square$

EXERCISES

14.1 Find, up to isomorphism, all abelian groups of order:

    a) 48;

    b) 72;

    c) 84;

    d) 450;

    e) 900.

14.2 Let X be a set and let $Y \subseteq X$. Let $G = (P(X), \triangle)$, and let A and B be the subgroups $(P(Y), \triangle)$ and $(P(X - Y), \triangle)$, respectively. Show that $G \cong A \times B$.

14.3 Characterize those finite abelian groups that are not isomorphic to the product of two nontrivial subgroups.

14.4 Let n be a positive integer. Show that every abelian group of order n is cyclic iff n is not divisible by the square of any prime.

14.5 Fill in the details in the classification of the groups of order 8.

14.6 Show that the conditions (i)–(iii) in the statement of Theorem 14.1 are equivalent to (i) and (ii)', where (ii)' asserts that every element of G has a *unique* representation in the form ab, with $a \in A$ and $b \in B$.

14.7 Let $G_1, \ldots, G_n$ be subgroups of G such that:

  i) $G_1, \ldots, G_n$ are all normal;
  ii) $G = G_1 G_2 \cdots G_n$, that is, every element of G can be written as $g_1 g_2 \cdots g_n$, with $g_i \in G_i$;
  iii) for $1 < i < n$, $G_i \cap G_1 G_2 \cdots G_{i-1} = \{e\}$.
     Show that $G \cong G_1 \times G_2 \times \cdots \times G_n$.

14.8 Show, by an example, that if we replace (iii) in Exercise 14.7 by the weaker condition $G_i \cap G_j = \{e\}$ for $i \neq j$, then G does not have to be isomorphic to $G_1 \times G_2 \times \cdots \times G_n$.

14.9 Let G, H, and K be finite abelian groups. Show that if $G \times H \cong G \times K$, then $H \cong K$.

14.10 Show, by example, that if we allow the group G in Exercise 14.9 to be infinite, then H need not be isomorphic to K. (Use an infinite direct product.)

14.11 Let G be an abelian group of order p^n, where p is a prime. An element $x \in G$ is said to be *of maximal order* if $o(x) \geq o(y)$ for all $y \in G$. Show that the only subgroup of G that contains all the elements of maximal order is G itself.

14.12 Let G and H be finite abelian groups such that for every integer n, G and H have the same number of elements of order n. Prove that $G \cong H$.

14.13 Let G be an abelian group of order p^n with invariants $p^{t_1} \geq p^{t_2} \geq \cdots \geq p^{t_r}$. Let H be a subgroup of G with invariants $p^{u_1} \geq p^{u_2} \geq \cdots \geq p^{u_s}$. Show that $s \leq r$ and $u_i \leq t_i$ for $1 \leq i \leq s$.

SYLOW THEOREMS

We have seen that for finite *abelian* groups, Lagrange's Theorem has a converse: if m divides $|G|$ and G is abelian, then G must have a subgroup of order m. Of course for general groups G this falls apart; A_4 is a group of order 12 with no subgroup of order 6. We are left wondering what can be salvaged in general; if G is a group and m divides $|G|$, then under what conditions *can* we assert that G must have a subgroup of order m?

Well, look at A_4. It *has* subgroups of order 2, 3, and 4, namely $\langle (1,2)(3,4) \rangle$, $\langle (1,2,3) \rangle$, and $\{e, (1,2)(3,4), (1,3)(2,4), (1,4)(2,3)\}$. At least on the basis of this (admittedly flimsy) evidence, we might suspect that the trouble comes when we "mix primes," i.e., when we try an m that is not just a power of some one prime. This suspicion can be borne out. For we shall prove in this section that if p^k divides $|G|$ for *any* finite group G, then G must have a subgroup of order p^k.

This assertion, together with some related facts, comprises what are known as the three *Sylow Theorems*, after the Norwegian mathematician Ludwig Sylow (1832–1918). We will state the three theorems together, then look at some examples and applications, and finally present the proofs.

We need to recall a couple of old notions before we can get started. If H is a subgroup of G and $g \in G$, then the set gHg^{-1} is called the **conjugate of H by g**. By Theorem 11.4, gHg^{-1} is a subgroup of G, with the same number of elements as H. If K is also a subgroup of G, we say that H and K are **conjugate** if $K = gHg^{-1}$ for some $g \in G$. Conjugacy is an equivalence relation on the set of all subgroups of G by the same proof as for conjugacy of elements. Finally, the **normalizer** of H in G is the subset

$$N(H) = \{ g \in G \mid gHg^{-1} = H \}.$$

By Ex/ cise 11.23, $N(H)$ is a subgroup of G. We have $H \subseteq N(H)$.

Example Let $G = S_3$, and let $H = \langle f \rangle$. Then $H \lhd G$, so $gHg^{-1} = H$ for every $g \in G$, and $N(H) = G$. On the other hand, if $H = \langle g \rangle$, then H is not normal, so $N(H)$ is a proper subgroup of S_3. Thus $|N(H)| = 1$, 2, or 3. But since $H \subseteq N(H)$ and $|H| = 2$, the only possibility is $|N(H)| = 2$. Thus $N(H) = H$.

We can now state the Sylow Theorems. If p is a prime such that p^n divides $|G|$ but p^{n+1} does not, then any subgroup of order p^n in G is called a **p-Sylow subgroup** of G.

THEOREM 15.1 (First Sylow Theorem) Let G be a finite group, p a prime.

i) If p^k divides $|G|$, then G has a subgroup of order p^k. In particular, G has a p-Sylow subgroup.

ii) Let H be any p-Sylow subgroup of G. If K is any subgroup of order p^k in G, then for some $g \in G$ we have $K \subseteq gHg^{-1}$. In particular, K is contained in some p-Sylow subgroup of G.

THEOREM 15.2 (Second Sylow Theorem) All p-Sylow subgroups of G are conjugate to each other. Consequently, a p-Sylow subgroup is normal iff it is the only p-Sylow subgroup.

THEOREM 15.3 (Third Sylow Theorem) Let H be any p-Sylow subgroup of G. Then the number of p-Sylow subgroups in G is $[G:N(H)]$. This number divides $|G|$ and has the form $1 + jp$ for some $j \geq 0$.

Examples Let $G = A_4$, a group of order 12. A 2-Sylow subgroup of G would be a subgroup of order 4, and

$$H = \{e, (1,2)(3,4), (1,3)(2,4), (1,4)(2,3)\}$$

is an example. As we have remarked before, all elements of $G - H$ are 3-cycles, hence have order 3. Thus H is the only 2-Sylow subgroup, and $H \lhd G$. In addition, all subgroups of order 2 or 4 in G are contained in H, illustrating part (ii) of the First Sylow Theorem.

A 3-Sylow subgroup of G would be a subgroup of order 3. If H is any such subgroup, then the number of 3-Sylow subgroups is $[G:N(H)]$. Since $H \subseteq N(H)$, we have

$$4 = [G:H] = [G:N(H)][N(H):H],$$

so the possibilities for $[G:N(H)]$ are 4, 2, and 1. But $[G:N(H)]$ must have the form $1 + 3j$, so 2 is ruled out. In fact, there are four 3-Sylow subgroups, namely $\langle (1,2,3) \rangle$, $\langle (1,2,4) \rangle$, $\langle (1,3,4) \rangle$, and $\langle (2,3,4) \rangle$. By Theorem 15.2, these subgroups are all conjugate to each other, and none of them is normal. For instance, we have

$$(1,2,4)\langle(1,2,3)\rangle(1,2,4)^{-1}=\langle(2,3,4)\rangle.$$

The Sylow Theorems are very useful in discussing the structure of finite groups. For example, we saw in Section 14 that a finite abelian group is a p-group iff $|G|=p^k$ for some k. We can now see that this holds for non-abelian G's as well.

THEOREM 15.4 Let G be a finite group, p a prime. Then G is a p-group iff $|G|$ is a power of p.

PROOF. If $|G|=p^k$, then clearly every element of G has order p^r, $0\leqslant r\leqslant k$. Conversely, if $|G|$ is not a power of p, there is some prime $q\neq p$ that divides $|G|$. By Theorem 15.1, G has a subgroup of order q, and therefore G has an element of order q, so that not every element of G has order a power of p. $\square$

As another application, we have

THEOREM 15.5 Let G be a group of order pq, where p and q are primes and $p<q$. Then if p does not divide $q-1$, G is cyclic.

For example, every group of order 35 is cyclic, so there is essentially only one group of order 35.

PROOF. Let P and Q be p-Sylow and q-Sylow subgroups of G, respectively. The number of p-Sylow subgroups is $[G:N(P)]$, which divides $[G:P]$, hence must be 1 or q. Since this number must have the form $1+jp$, it cannot be q, because $q=1+jp$ would give us $p|(q-1)$, contrary to our assumption. Thus P is the only p-Sylow subgroup, and $P\triangleleft G$. The number of q-Sylow subgroups is $[G:N(Q)]$, hence must be 1 or p. It can't be p, because $p=1+kq$ would make $p>q$. Thus Q is the only q-Sylow subgroup, and $Q\triangleleft G$.

So both P and Q are normal. It is obvious that $P\cap Q=\{e\}$, and therefore

$$|PQ|=\frac{|P|\cdot|Q|}{|P\cap Q|}=\frac{pq}{1}=pq,$$

so $PQ=G$. Thus $G\cong P\times Q$, and since $(p,q)=1$, G is cyclic. $\square$

What happens to Theorem 15.5 if p does divide $q-1$? See Exercises 15.17 and 15.18.

The Sylow Theorems can be very incisive tools for analyzing the structure of the groups of some given specific order, as the following examples illustrate.

Examples

1. A group G is called **simple** if its only normal subgroups are $\{e\}$ and G. For instance, any group of prime order is simple, and it can be shown that A_n is simple, for $n \geqslant 5$.

We assert that no group G of order 28 is simple. For let H be a 7-Sylow subgroup of G, and consider the number of 7-Sylow subgroups of G. This number is $[G:N(H)]$, hence must divide 4; but it is also of the form $1 + 7j$. Clearly, then, there is only one 7-Sylow subgroup, so $H \lhd G$.

2. A group of order 24 cannot be simple. In fact, G must have a normal subgroup of order either 4 or 8. To see this, note first that the number of 2-Sylow subgroups has the form $1 + 2j$ and divides 3, hence is either 1 or 3. Thus we are not assured of the existence of a normal subgroup of order 8. However, let H be some 2-Sylow subgroup. Then by Exercise 13.29, there is a normal subgroup N of G such that $N \subseteq H$ and $[G:N]$ divides $[G:H]! = 3! = 6$. If $|N| = 1$ or 2, then $[G:N] = 24$ or 12, and neither of these divides 6. Hence $|N| = 4$ or 8.

3. We will determine all groups of order 1225. Since $1225 = 5^2 \cdot 7^2$, we know that if $|G| = 1225$ then G has Sylow subgroups A and B of orders 25 and 49, respectively. The number of 5-Sylow subgroups has the form $1 + 5j$ and divides 49, hence is 1. The number of 7-Sylow subgroups has the form $1 + 7k$ and divides 25, hence is 1. Thus A and B are both normal, $A \cap B = \{e\}$, and $AB = G$, since AB has order $(|A| \cdot |B|)/1 = 25 \cdot 49 = |G|$. We conclude that $G \cong A \times B$, and since both A and B are abelian (Exercise 10.25), we see that G is abelian.

Thus any group of order 1225 is isomorphic to one of

$$\mathbb{Z}_{25} \times \mathbb{Z}_{49}, \quad \mathbb{Z}_{25} \times \mathbb{Z}_7 \times \mathbb{Z}_7, \quad \mathbb{Z}_5 \times \mathbb{Z}_5 \times \mathbb{Z}_{49}, \quad \text{or} \quad \mathbb{Z}_5 \times \mathbb{Z}_5 \times \mathbb{Z}_7 \times \mathbb{Z}_7.$$

4. We will determine all groups G of order 30.

First of all, since $30 = 2 \cdot 3 \cdot 5$, we know that there are Sylow subgroups A, B, and C of order 2, 3, and 5, respectively. The number of 5-Sylow subgroups is $[G:N(C)]$, hence must divide 6. But this number is also of the form $1 + 5j$, hence the number of 5-Sylow subgroups is either 1 or 6. Similarly, the number of 3-Sylow subgroups has the form $1 + 3k$ and divides 10, so must be either 1 or 10.

Now suppose there were six 5-Sylow subgroups and ten 3-Sylow subgroups. Any two distinct 5-Sylow subgroups must have trivial intersection (since they both have order 5), so all six together would give us $6 \cdot 4 = 24$ elements of order 5 in G. Similarly, the 3-Sylow subgroups would give us 20 elements of order 3 in G. Thus we would have $|G| \geqslant 44$, which is nonsense.

Thus, either there is only one 3-Sylow subgroup, or there is only one 5-Sylow subgroup. In other words, one of B, C must be normal. Thus BC is a subgroup of G, of order $(|B| \cdot |C|)/(|B \cap C|) = 15$. By Theorem 15.5, BC is cyclic; say $BC = \langle x \rangle$.

Since $\langle x \rangle$ has index 2, it is normal in G. If we let $A = \langle y \rangle$, then

$$G = \langle x \rangle \langle y \rangle,$$

since $\langle x \rangle \langle y \rangle$ has order 30. We must have $yxy^{-1} = x^t$ for some integer t, and if we knew the value of t, then the structure of G would be determined. (We would know that $yx^n y^{-1} = x^{nt}$ for every integer n, hence $yx^n = x^{nt}y$. This tells us explicitly how to multiply any two elements in $\langle x \rangle \langle y \rangle$.)

Now yxy^{-1} must have order 15, because x does, and therefore $(t, 15) = 1$, so $t = 1, 2, 4, 7, 8, 11, 13,$ or 14. Moreover, we have

$$y(yxy^{-1})y^{-1} = yx^t y^{-1} = (yxy^{-1})^t = (x^t)^t = x^{t^2},$$

that is, $x = x^{t^2}$, so $x^{t^2-1} = e$, and thus 15 divides $t^2 - 1$. This rules out $t = 2, 7, 8,$ and 13, so there are at most four possibilities for t, and hence at most four nonisomorphic groups of order 30.

Actually, there are precisely four, because $\mathbb{Z}_{30}$, $S_3 \times \mathbb{Z}_5$, $\mathbb{Z}_3 \times D_5$, and D_{15} (see Exercises 8.12 and 8.20) are pairwise nonisomorphic. For instance, their centers have orders 30, 5, 3, and 1, respectively.

Examples 3 and 4 might start you wondering what the number of nonisomorphic groups of order n is for various n's. The following table gives the answers for $n \leqslant 23$; there is no known formula for computing the answers in general.

Order of group	1	2	3	4	5	6	7	8	9	10	11	12	13	14	15	16	17	18	19	20	21	22	23
Number of groups	1	1	1	2	1	2	1	5	2	2	1	5	1	2	1	14	1	5	1	5	2	2	1

We now turn to the proofs of the Sylow Theorems. There are several ways of approaching them; we have chosen the route that we think is the cleanest and the most illustrative of things we have talked about before. To be honest about it, though, we have chosen the following proofs mostly because we think they are beautiful.

The proof of part (i) of the First Sylow Theorem brings back our method of "quotient groups and induction." We let G be a finite group, and proceed by induction on $|G|$. If $|G| = 2$, the result is trivial. Now assume the statement is true for all groups of order less than $|G|$, and suppose p^k divides $|G|$. If G has a proper subgroup H whose index is not divisible by p, then p^k divides $|H|$, so by the inductive hypothesis H has a subgroup of order p^k, which is of course also a subgroup of G. Thus we may as well assume that $p | [G:H]$, for every proper subgroup H of G.

From this it follows, via the class equation, that $Z(G)$ has a subgroup of order p. For if we choose $\{g_1,\ldots,g_r\}$ to consist of one representative from each conjugacy class in G that has at least two elements, then

$$|G| = |Z(G)| + [G:Z(g_1)] + \cdots + [G:Z(g_r)],$$

and each $Z(g_j)$ is a proper subgroup since $g_j \notin Z(G)$. Hence each term $[G:Z(g_j)]$ is divisible by p, and since p divides $|G|$ this implies that p divides $|Z(G)|$. Therefore $Z(G)$ is a finite abelian group whose order is divisible by p, so $Z(G)$ has a subgroup A of order p, either by Theorem 11.7 or by Corollary 14.5.

Now $A \lhd G$ since $A \subseteq Z(G)$, and therefore G/A is a group of order $|G|/p$. If p^k divides $|G|$, then p^{k-1} divides $|G/A|$, so by the inductive hypothesis G/A has a subgroup J of order p^{k-1}. If $\rho: G \rightarrow G/A$ is the canonical homomorphism, then

$$\rho^{-1}(J)/A \cong J,$$

so $\rho^{-1}(J)$ has order p^k and part (i) is proved.

To prove part (ii), we want to start with some given subgroup K of order p^k in G and some given p-Sylow subgroup H, and find a conjugate H^* of H such that $K \subseteq H^*$. We proceed in two steps:

LEMMA 15.6 With K, H as just indicated, there is a conjugate H^* of H such that $K \subseteq N(H^*)$.

LEMMA 15.7 With K, H as indicated, $K \subseteq N(H^*)$ implies $K \subseteq H^*$.

Much of what we do from this point on will be strongly reminiscent of the development of the class equation in Section 10, because we shall proceed by counting conjugates. Specifically, the claim that $K \subseteq N(H^*)$ in Lemma 15.6 just says that $kH^*k^{-1} = H^*$ for every $k \in K$, or, in other words, the number of distinct conjugates of H^* by elements of K is 1. We count conjugates in Lemma 15.8, which is a cousin of Lemma 10.8.

LEMMA 15.8 Let G be a group, and let K and L be subgroups of G. Then the number of distinct conjugates of L by elements of K is $[K : K \cap N(L)]$. In particular, the number of distinct conjugates of L in G is $[G : N(L)]$.

PROOF. Let $x, y \in K$. Then $xLx^{-1} = yLy^{-1}$ iff $L = x^{-1}yLy^{-1}x$ iff $x^{-1}y \in N(L)$ iff $x^{-1}y \in K \cap N(L)$. Thus if $x, y \in K$, then $xLx^{-1} = yLy^{-1}$ iff x and y are in the same left coset of $K \cap N(L)$ in K. Therefore the mapping

$$xLx^{-1} \rightarrow x(K \cap N(L))$$

establishes a one-to-one correspondence between the conjugates of L by elements of K and the left cosets of $K \cap N(L)$ in K. This proves the lemma.
□

Example Let $G = D_4 = \{e, f, f^2, f^3, g, fg, f^2g, f^3g\}$, let $L = \langle g \rangle = \{e, g\}$, and let $K = \langle f \rangle$. We know that $N(L)$ contains L, and it also contains f^2, since $f^2 \in Z(D_4)$. Thus $\{e, f^2, g, f^2g\} \subseteq N(L)$. Since L is not normal, $|N(L)| < 4$, so $N(L) = \{e, f^2, g, f^2g\}$. This gives us

$$[K : K \cap N(L)] = [\langle f \rangle : \langle f^2 \rangle] = 2,$$

so there should be exactly two conjugates of L by elements of K. Indeed this is the case, since

$$e \langle g \rangle e^{-1} = f^2 \langle g \rangle f^{-2} = \langle g \rangle,$$

and

$$f \langle g \rangle f^{-1} = f^3 \langle g \rangle f^{-3} = \langle f^2g \rangle.$$

PROOF OF LEMMA 15.6. Let K be a subgroup of order p^k in G, and let H be a p-Sylow subgroup. If $H_1, H_2, \ldots, H_m$ are all the conjugates of H in G, we wish to show that $K \subseteq N(H_i)$ for some i, that is, $[K : K \cap N(H_i)] = 1$. Since $[K : K \cap N(H_i)]$ must be among $1, p, p^2, \ldots, p^k$ (because $|K| = p^k$), all we have to show is that some $[K : K \cap N(H_i)]$ is not divisible by p.

Consider the equivalence relation R on the set $\{H_1, H_2, \ldots, H_m\}$ given by

$$H_i R H_j \quad \text{iff there exists } x \in K \text{ such that} \quad H_i = xH_jx^{-1}.$$

The equivalence class of any H_i consists of all the conjugates of H_i by elements of K, and therefore contains $[K : K \cap N(H_i)]$ many elements. If every $[K : K \cap N(H_i)]$ were divisible by p, then the sum of the orders of the equivalence classes would be divisible by p, that is, p would divide m. But p does *not* divide m, because by Lemma 15.8, $m = [G : N(H)]$; if $p | [G : N(H)]$, then $p | [G : H]$, contradicting the fact that $|H|$ is the highest power of p that divides $|G|$. $\square$

PROOF OF LEMMA 15.7. Suppose K is a subgroup of order p^k, H is a p-Sylow subgroup of order p^n, and $K \subseteq N(H^*)$ for some conjugate H^* of H. If $K \not\subseteq H^*$, then $K \cap H^*$ is a proper subgroup of K, of order p^j, for some $j < k$. Then since $H^* \lhd N(H^*)$, KH^* is a subgroup of $N(H^*)$, of order

$$\frac{|K| \cdot |H^*|}{|K \cap H^*|} = \frac{p^k p^n}{p^j} = p^n p^{k-j} > p^n.$$

This contradicts the fact that H is a p-Sylow subgroup of G. Thus Lemma 15.7 is proved and, with it, the First Sylow Theorem. $\square$

PROOF OF THE SECOND SYLOW THEOREM. Easy: If K and H are p-Sylow subgroups, then by part (ii) of the First Sylow Theorem, we have $K \subseteq gHg^{-1}$ for some $g \in G$. But K and gHg^{-1} have the same order, so $K = gHg^{-1}$, and K and H are conjugate.

Next, if a p-Sylow subgroup H is normal and K is any p-Sylow subgroup, then the fact that $K = gHg^{-1}$ for some g means that $K = H$. Thus H is the only p-Sylow subgroup. Conversely, if H is the only p-Sylow subgroup, then H is normal by Corollary 11.5. $\square$

PROOF OF THE THIRD SYLOW THEOREM. Let H be a p-Sylow subgroup. By the Second Sylow Theorem, the number of p-Sylow subgroups is the number of conjugates of H in G, and this is $[G:N(H)]$ by Lemma 15.8. Finally, if $H_1, \ldots, H_m$ are all the p-Sylow subgroups of G, we want to show that $m = 1 + jp$ for some $j \geq 0$. If K is any one of the p-Sylow subgroups, then as in the proof of Theorem 15.1(ii), we have

$$m = [K : K \cap N(H_1)] + \cdots + [K : K \cap N(H_m)],$$

where each term in the sum is among $1, p, p^2, \ldots, p^n = |K|$. To finish the proof, it will suffice to show that exactly one term in the sum is 1. But what does $[K : K \cap N(H_i)] = 1$ mean? It means that $K \subseteq N(H_i)$, which, by Lemma 15.7, means that $K \subseteq H_i$. Since $|K| = |H_i|$, this means that $K = H_i$. Thus $[K : K \cap N(H_i)] = 1$ iff $H_i = K$, and the proof is complete. $\square$

EXERCISES

15.1 Let H be a normal subgroup of a finite group G and suppose $|H| = p^k$, where p is a prime. Show that H is contained in every p-Sylow subgroup of G.

15.2 Let H be a p-Sylow subgroup of G. Show that H is the only p-Sylow subgroup of G contained in $N(H)$.

15.3 Suppose that $K \triangleleft H \triangleleft G$ and K is a p-Sylow subgroup of H. Show that $K \triangleleft G$.

15.4 Assume that all the Sylow subgroups of G are normal. Show that G is isomorphic to their direct product.

15.5 Show that no group of order 56 is simple.

15.6 Show that no group of order 200 is simple.

15.7 In Exercise 10.14, we showed that there are essentially only two groups of order 6. Give a slicker proof of this result, by using the Sylow Theorems.

15.8 Find all groups of order 4225.

15.9 Show that if $|G| = 66$, then G has at least three normal subgroups other than $\{e\}$ and G.

15.10 Show that every group of order 255 is abelian.

15.11 Show that every group of order 455 is abelian.

15.12 Complete the classification of groups of order 30 by verifying the parts that were left as exercises.

15.13 Prove that there are at most five pairwise nonisomorphic groups of order 18.

15.14 Give an example of a group of order 24 that has no normal subgroup of order 8.

15.15 Let G be a group of order p^2q, where p and q are distinct primes. Show that G has either a normal p-Sylow subgroup or a normal q-Sylow subgroup.

15.16 Let G be a group of order p^2q, where p and q are primes such that $q < p$ and q does not divide $p^2 - 1$. Show that G is abelian.

15.17 Carry out the following steps to show that if p and q are primes such that p divides $q - 1$, then there exists a nonabelian group of order pq.

a) Show that there exists an integer t, $1 \leqslant t \leqslant q - 1$, such that $t \not\equiv 1 \pmod{q}$, but $t^p \equiv 1 \pmod{q}$. [*Hint:* Consider the group $(\mathbb{Z}_q - \{0\}, \odot)$.]

b) Let G be the set of all ordered pairs (x, y), where $x \in (\mathbb{Z}_q, \oplus)$ and $y \in (\mathbb{Z}_p, \oplus)$. Let t be an integer as in (a) and define

$$(x, y) * (u, v) = (x \oplus t^y u, y \oplus v),$$

where the first entry is computed mod q and the second mod p. Show that G forms a group of order pq under this operation. G is an example of what is called a **semidirect product**.

c) Show that $(0, 1)(1, 0)(0, 1)^{-1} = (t, 0)$, and therefore $\langle (1, 0) \rangle$ is normal in G, but G is not abelian.

15.18 Let p, q be primes such that p divides $q - 1$. Assume the following fact: If t is an integer which satisfies the conditions $t \not\equiv 1 \pmod{q}$ and $t^p \equiv 1 \pmod{q}$, then any integer which satisfies these conditions must be congruent $\pmod{q}$ to one of $t, t^2, \ldots, t^{p-1}$. Use this to prove that all nonabelian groups of order pq are isomorphic to each other. [*Suggestions:* Note that the q-Sylow subgroup Q of any such group must be normal. Let $Q = \langle a \rangle$ and take $b \in G - Q$. Then $\langle a \rangle \langle b \rangle = G$ and $bab^{-1} = a^t$ for some t. Show that $b^p a b^{-p} = a^{t^p}$, and therefore $t^p \equiv 1 \pmod{q}$, although $t \not\equiv 1 \pmod{q}$. Note that if we choose a different generator for $\langle b \rangle$—for example, b^j, $2 < j < p - 1$—then $b^j a (b^j)^{-1} = a^{t^j}$.]

15.19 Let H be a p-Sylow subgroup of G. Show that $N(N(H)) = N(H)$. (*Suggestion:* Use Exercise 15.2.)

15.20 Let $|G| = p^n$ and let H be a subgroup of G such that $|H| = p^m$, where $m \leqslant n$. Show that if $m < k < n$, then there exists a subgroup K of G such that $|K| = p^k$ and $H \subseteq K$. (*Suggestion:* Use the fact that, by the class equation, $Z(G)$ has a subgroup of order p.)

15.21 Suppose that $|G| = p^n$, $H \triangleleft G$, and $|H| = p$. Prove that $H \subseteq Z(G)$.

15.22 Let H be a p-Sylow subgroup of G and let $K \triangleleft G$. Show that $H \cap K$ is a p-Sylow subgroup of K.

SECTION 16

RINGS

Up to now we have been studying sets with a single binary operation defined on them. For example, we have encountered groups such as $(\mathbb{Q}, +)$ and $(\mathbb{Q} - \{0\}, \cdot)$. But of course there are times in real life when one considers both addition and multiplication simultaneously, for instance on $\mathbb{Q}$, or more basically on $\mathbb{Z}$. We will now consider an abstract notion designed to capture the essence of such situations where two operations interact with each other.

What *is* the essence of the situation for addition and multiplication on $\mathbb{Z}$, for example? If we just look at addition, then we have an abelian group $(\mathbb{Z}, +)$. If we concentrate on multiplication, we have an associative commutative binary operation. There happens to be an identity element for $\cdot$, but most elements fail to have inverses. Finally, if we consider both operations at once, then the most salient point is that they are connected by the **distributive laws**:

$$a(b + c) = ab + ac \quad \text{and} \quad (b + c)a = ba + ca.$$

During the nineteenth century, number theorists worked with systems more inclusive than $\mathbb{Z}$ which satisfied the same properties with respect to $+$ and $\cdot$. One motivation for their efforts was the hope that by considering such systems, one might answer questions about $\mathbb{Z}$ that could not be answered by thinking in terms of $\mathbb{Z}$ alone. Although this hope was only partially realized (questions about $\mathbb{Z}$ can be hard!), a great deal was accomplished, and moreover, the groundwork was laid for the development of an abstract theory in the twentieth century.

The abstract concept which emerged is that of a *ring*. A ring has all the properties of $\mathbb{Z}$ indicated above, except that in order to achieve greater generality, one does not require that there exist an identity element for multiplication, nor that multiplication be commutative. (A good deal of work has also been done on systems for which multiplication fails even to be associative, but we will not consider such nonassociative rings.)

In writing down the axioms for a ring, we should perhaps use symbols such as $*$ and $\square$ to denote the two operations involved, to emphasize that they do not have to be ordinary addition and multiplication of numbers. However, we shall just use $+$ and $\cdot$, for simplicity. You are experienced enough by this time to keep in mind that we are just talking about two binary operations, even though we denote them by $+$ and $\cdot$ and call them addition and multiplication.

DEFINITION Suppose that R is a set and $+$ and $\cdot$ are two binary operations on R. Suppose further that:

i) $(R, +)$ is an abelian group,

ii) $\cdot$ is associative, and

iii) the distributive laws hold, i.e.,

$$r_1 \cdot (r_2 + r_3) = r_1 \cdot r_2 + r_1 \cdot r_3 \quad \text{and} \quad (r_2 + r_3) \cdot r_1 = r_2 \cdot r_1 + r_3 \cdot r_1,$$

for all r_1, r_2, r_3 in R.

Then R, together with the binary operations $+$ and $\cdot$, is called a **ring**. We denote it by $(R, +, \cdot)$, or R for short.

The two distributive laws are referred to as the *left* and *right* distributive laws, respectively. Of course, if $\cdot$ happens to be commutative, then these two laws say the same thing.

A ring for which $\cdot$ is commutative is called a **commutative ring**. In general, the addition on a ring has already been assumed to be pretty nice, and one gets more special rings by imposing more assumptions on the multiplication.

The **additive identity element** of R, i.e., the identity element for $(R, +)$, is denoted by 0, or 0_R. If there happens to be an identity element for $\cdot$, then it is an easy matter to see that there is only one such; it is called the **multiplicative identity element** or the **unity** of R, and is denoted by 1, or 1_R. A ring that possesses a unity is called (what else?) a **ring with unity**.

Examples

1. $(\mathbb{Z}, +, \cdot)$ is a commutative ring with unity, as are $(\mathbb{Q}, +, \cdot)$ and $(\mathbb{R}, +, \cdot)$. Here $+$ and $\cdot$ denote ordinary addition and multiplication.

2. $(2\mathbb{Z}, +, \cdot)$ is a commutative ring, but not a ring with unity.

3. Let R be the set of all real numbers that can be written in the form $a + b\sqrt{2}$, where $a, b \in \mathbb{Z}$. It is clear that the sum of two elements of R is in R, and if $a + b\sqrt{2}$ and $c + d\sqrt{2}$ are in R, then their product

$$(ac + 2bd) + (ad + bc)\sqrt{2}$$

is in R too. R is a commutative ring with unity under ordinary addition and multiplication.

4. Let $R = \{0, 1, 2, \ldots, n - 1\}$, and let $\oplus$ and $\odot$ denote addition and multiplication modulo n on R, that is,

$$a \oplus b = \overline{a + b} \quad \text{and} \quad a \odot b = \overline{a \cdot b},$$

where $\overline{}$ denotes remainders modulo n. Then $(R, \oplus)$ is the familiar group $(\mathbb{Z}_n, \oplus)$, and we claim that $(R, \oplus, \odot)$ is a commutative ring with unity, which we will denote by $(\mathbb{Z}_n, \oplus, \odot)$. In fact, the proofs of associativity for $\odot$ and of the distributive laws are very similar to the proof we gave for the associativity of $\oplus$ on $\mathbb{Z}_n$ in Section 2 (Exercise 16.8). $(\mathbb{Z}_n, \oplus, \odot)$ is commutative because

$$a \odot b = \overline{a \cdot b} = \overline{b \cdot a} = b \odot a.$$

The multiplicative identity element is 1.

The rings $(\mathbb{Z}_n, \oplus, \odot)$ are interesting at this point because they already begin to display behavior quite different from that of the prototype example $(\mathbb{Z}, +, \cdot)$. For instance, look at $(\mathbb{Z}_6, \oplus, \odot)$. Here $2 \odot 3 = 0$ although neither 2 nor 3 is 0. Things like that certainly don't happen in $\mathbb{Z}$. To see something even stranger, look at $(\mathbb{Z}_8, \oplus, \odot)$. Here $2^3 (= 2 \odot 2 \odot 2) = 0$, so a power of a nonzero element can be 0.

We introduce some terminology in order to deal with such situations. An element $a \in R$ is called a **zero-divisor** if there exists an element $b \neq 0$ such that either

$$ab = 0 \quad \text{or} \quad ba = 0.$$

a is called **nilpotent** if there exists some positive integer n such that $a^n = 0$. (Here a^n means a multiplied by itself n times.) Thus in $\mathbb{Z}_6$, 2 is a zero-divisor, and so is 3 (and 0 and 4). In $\mathbb{Z}_8$, 2 is nilpotent, as are 0, 4, and 6.

At the opposite extreme from these badly behaved elements are those called *units*. Suppose R is a ring with unity. Then $a \in R$ is called a **unit** if there exists an element $b \in R$ such that

$$ab = ba = 1.$$

Note right away that "unit" and "unity" are not the same thing! The unity is a unit because $1 \cdot 1 = 1$, but there may be many units other than 1. For instance, in $\mathbb{Z}_8$, 1, 3, 5, and 7 are all units, since $1^2 = 3^2 = 5^2 = 7^2 = 1$. In $\mathbb{Z}_{10}$, 1, 3, 7, and 9 are units because $1^2 = 9^2 = 1$ and $3 \odot 7 = 7 \odot 3 = 1$.

We shall see below that a unit can never be a zero-divisor. The general situation for $(\mathbb{Z}_n, \oplus, \odot)$ is that $a \in \mathbb{Z}_n$ is a unit iff $(a, n) = 1$, and every element is either a unit or a zero-divisor (see Exercise 16.9). On the other hand, some rings contain elements that are neither units nor zero-divisors. Can you give an example?

It is easy to see that if a is a unit, then there is only one b such that $ab = ba = 1$. We call b the **multiplicative inverse of a**, and denote it by a^{-1}.

Examples (continued)

5. Let $\mathbb{R} \oplus \mathbb{R}$ be the set of all ordered pairs (a, b) of real numbers, with addition and multiplication defined componentwise:

$$(a,b) + (c,d) = (a+c, b+d) \quad \text{and} \quad (a,b) \cdot (c,d) = (ac, bd).$$

Then $\mathbb{R} \oplus \mathbb{R}$ is a commutative ring with unity. An element (a, b) in $\mathbb{R} \oplus \mathbb{R}$ is a zero-divisor iff at least one of a, b is 0, and it is a unit iff neither of a, b is 0. For example, the equation

$$(0,3) \cdot (1,0) = (0,0)$$

shows that both $(0,3)$ and $(1,0)$ are zero-divisors, while

$$(5,6) \cdot \left(\frac{1}{5}, \frac{1}{6} \right) = (1,1)$$

shows that both $(5,6)$ and $\left(\frac{1}{5}, \frac{1}{6} \right)$ are units. In this ring it is true that an element is a zero-divisor iff it is not a unit.

The ring $\mathbb{R} \oplus \mathbb{R}$ is called a *direct sum*. It is common practice, in ring theory, to speak of "direct sums" rather than "direct products." The additive terminology is also commonly used in discussing abelian groups.

6. Generalizing the previous example, let $R_1, R_2, \ldots, R_n$ be rings. Then their **direct sum** $R_1 \oplus R_2 \oplus \cdots \oplus R_n$ is the ring whose elements are all n-tuples $(r_1, r_2, \ldots, r_n)$, with $r_i \in R_i$, under componentwise addition and multiplication. The direct sum is commutative iff each summand R_i is, and it has a multiplicative identity iff each R_i has one.

7. Let R be the set of all real-valued functions defined on $\mathbb{R}$, under addition and multiplication of functions. R is a commutative ring with unity. An element f in R is a zero-divisor iff $f(x) = 0$ for at least one $x \in \mathbb{R}$, and it is a unit iff $f(x) \neq 0$ for all $x \in \mathbb{R}$. Here, too, an element is a zero-divisor iff it is not a unit.

Which elements of R are nilpotent?

8. Let $M_2(\mathbb{R})$ denote the set of *all* 2×2 matrices with real entries, under addition and multiplication of matrices. (Addition means adding corresponding entries.) Then $M_2(\mathbb{R})$ is a noncommutative ring with unity. For example, the left distributive law requires that

$$\begin{pmatrix} a & b \\ c & d \end{pmatrix}\left[\begin{pmatrix} e & f \\ g & h \end{pmatrix}+\begin{pmatrix} i & j \\ k & l \end{pmatrix}\right]=\begin{pmatrix} a & b \\ c & d \end{pmatrix}\begin{pmatrix} e & f \\ g & h \end{pmatrix}+\begin{pmatrix} a & b \\ c & d \end{pmatrix}\begin{pmatrix} i & j \\ k & l \end{pmatrix},$$

and this is easily checked.

Note that there exist elements A, B in $M_2(\mathbb{R})$ such that $AB = \begin{pmatrix} 0 & 0 \\ 0 & 0 \end{pmatrix}$ but $BA \neq \begin{pmatrix} 0 & 0 \\ 0 & 0 \end{pmatrix}$. For instance, we can take

$$A = \begin{pmatrix} 1 & 0 \\ 1 & 0 \end{pmatrix} \quad \text{and} \quad B = \begin{pmatrix} 0 & 0 \\ 0 & 1 \end{pmatrix}.$$

This example explains why, in the definition of zero-divisor, we said "$ab = 0$ or $ba = 0$."

Note also that this ring contains nonzero nilpotent elements. For example,

$$\begin{pmatrix} 0 & 1 \\ 0 & 0 \end{pmatrix}^2 = \begin{pmatrix} 0 & 0 \\ 0 & 0 \end{pmatrix}.$$

9. Let $(G, +)$ be any abelian group, and denote the identity element of G by 0. Define a multiplication on G by declaring $a \cdot b = 0$ for all $a, b \in G$. It is then easy to check that $(G, +, \cdot)$ is a ring. We call this the **ring on G with trivial multiplication.**

If, in particular, we start with $G = \{0\}$, the trivial group, then we get a ring with one element. It is called the **trivial ring**, and is rather annoying. For example, 0 satisfies the definition of a multiplicative identity element ($0 \cdot x = x \cdot 0 = x$ for all x, right?), so 0 is 1. That should make you cringe, but there is some comfort in noting that this anomaly can only occur in the trivial ring: If R is a ring with unity and R has more than one element, then $0 \neq 1$ in R. In order to prove this, we need some basic information.

We frequently denote multiplication in a ring by juxtaposition, writing ab rather than $a \cdot b$.

THEOREM 16.1 Let R be a ring, and let a, b be elements of R. Then:

i) $a \cdot 0 = 0 \cdot a = 0$;

ii) $a(-b) = (-a)b = -(ab)$;

iii) $(-a)(-b) = ab$;

iv) $m(ab) = (ma)b = a(mb)$ for any integer m;

v) $mn(ab) = (ma)(nb)$ for any integers m and n.

PROOF. i) To show that $a \cdot 0 = 0$ it is enough to show that

$$a \cdot 0 + a \cdot 0 = a \cdot 0,$$

for then we can add $-(a \cdot 0)$ to both sides. But

$$a \cdot 0 + a \cdot 0 = a \cdot (0+0) = a \cdot 0.$$

Similar reasoning works for $0 \cdot a$.

ii) To show that $a(-b) = -(ab)$, it suffices to show that $a(-b) + ab = 0$. But

$$a(-b) + ab = a(-b+b) = a \cdot 0 = 0$$

by (i). A similar argument, using right distributivity instead of left, shows that $(-a)b = -(ab)$.

iii) Replacing a by $-a$ in the first equality of (ii), we obtain

$$(-a)(-b) = [-(-a)]b = ab.$$

iv) Exercise.

v) Exercise. $\square$

COROLLARY 16.2 Let R be a nontrivial ring with unity. Then $0 \neq 1$ in R.

PROOF. Since R is nontrivial, we can pick $a \neq 0$ in R. Then if $0 = 1$ we get $a \cdot 0 = a \cdot 1$, that is, $0 = a$, a contradiction. $\square$

COROLLARY 16.3 Let R be a ring with unity, and let $u \in R$ be a unit. Then u is not a zero-divisor in R.

PROOF. We must show that if r is an element of R such that $ur = 0$ or $ru = 0$, then $r = 0$. Now if $ur = 0$, then

$$u^{-1}(ur) = u^{-1}(0) = 0,$$

that is,

$$r = 0.$$

A similar argument works if $ru = 0$. $\square$

The next corollary is technical, but useful.

COROLLARY 16.4 If b and c are elements of a ring R, define $b - c$ to mean $b + (-c)$. Then for any $a \in R$, we have

$$a(b-c) = ab - ac, \quad \text{and} \quad (b-c)a = ba - ca.$$

PROOF.

$$a(b-c) = a(b+(-c)) = ab + a(-c) = ab + [-(ac)] = ab - ac.$$

Likewise for the second equality. $\square$

We mentioned above that one gets nicer and nicer rings by imposing more and more assumptions on the multiplication. For example, we get rings that behave somewhat like $(\mathbb{Z}, \oplus, \odot)$ by adding the assumptions indicated in the following

DEFINITION An **integral domain** (or just **domain**, for short) is a commutative ring with unity in which $1 \neq 0$ and there are no nonzero zero-divisors.

Thus $\mathbb{Z}$, $\mathbb{Q}$, and $\mathbb{R}$ are all domains. Another example is the ring introduced in Example 3 above, consisting of all real numbers of the form $a + b\sqrt{2}$, with $a, b \in \mathbb{Z}$.

The following simple observation yields an alternative characterization of integral domains.

THEOREM 16.5 Let R be a ring and let $a, b, c \in R$. Assume that a is not a zero-divisor. Then if $ab = ac$, we have $b = c$.

PROOF. From $ab = ac$ we get $ab - ac = 0$, so $a(b - c) = 0$. Since a is not a zero-divisor, this means $b - c = 0$, so $b = c$. $\square$

COROLLARY 16.6 Let R be a commutative ring with unity $1 \neq 0$. Then R is an integral domain iff whenever $a, b, c \in R$ satisfy $ab = ac$ and $a \neq 0$, we have $b = c$.

PROOF. Assume that R is a domain. Let $a, b, c \in R$, $a \neq 0$, and suppose that $ab = ac$. Then, since a is not a zero-divisor, the theorem implies that $b = c$.

Conversely, suppose that R is not an integral domain. We will find $a, b, c \in R$, $a \neq 0$, such that

$$ab = ac \quad \text{but} \quad b \neq c.$$

In fact, we know that there is a nonzero zero-divisor $a \in R$. Let $b \in R$ be such that $b \neq 0$ and $ab = 0$. Then we have

$$a \cdot b = a \cdot 0, \quad \text{but} \quad b \neq 0. \quad \square$$

Finally, we consider rings with unity in which every nonzero element has a multiplicative inverse.

DEFINITIONS R is called a **division ring** if R has a unity $1 \neq 0$ and every nonzero element of R is a unit. A commutative division ring is called a **field**.

Another way of saying that R is a division ring is to say that the set $R - \{0\}$ forms a group under multiplication. Saying that R is a field amounts to saying that this group is abelian.

Familiar examples of fields include $\mathbb{Q}$, $\mathbb{R}$, and the complex numbers $\mathbb{C}$. A less familiar example is $(\mathbb{Z}_p, \oplus, \odot)$, where p is a prime number. $\mathbb{Z}_p$ is a field because each $r \in \{1, 2, \ldots, p - 1\}$ satisfies $(r, p) = 1$ and is therefore a unit in $\mathbb{Z}_p$ (see Exercise 16.9).

Examples of division rings that are *not* fields are a little harder to come by. In fact, a celebrated theorem of J. H. M. Wedderburn asserts that there *aren't* any *finite* examples: Every finite division ring is necessarily a field. Proving this here would take us too far afield (so to speak), so we shall content ourselves with a much simpler result. We shall see an infinite division ring that is not a field in Section 17.

We obtain our easy substitute for Wedderburn's Theorem by replacing "division ring" by "integral domain."

THEOREM 16.7 Every finite integral domain is a field.

PROOF. Let R be a finite domain. Then R is commutative, and $1 \neq 0$ in R. We must show that if $r \in R$, $r \neq 0$, then r has a multiplicative inverse in R.

Since R is finite we can list its elements as $r_1, r_2, \ldots, r_n$. Consider the elements

$$rr_1, rr_2, rr_3, \ldots, rr_n.$$

Since $r \neq 0$, these are all distinct by Corollary 16.6. Since they are all in R and R has only n elements altogether, they must account for all the elements in R. In particular, one of them is 1, so

$$rr_i (= r_i r) = 1$$

for some i, and r is a unit. $\square$

This proof will look familiar to you if you worked Exercise 3.15.

Of course, there exist infinite domains that are not fields—$\mathbb{Z}$, for instance. On the other hand, every field, finite or infinite, is a domain, because units are not zero-divisors. Thus the notions of "domain" and "field" coincide for finite rings, but, in general, "field" is stronger.

EXERCISES

16.1 Let R be a ring with unity 1_R. Show that $(-1_R)a = -a$ for all $a \in R$.

16.2 a) If $r * s = 2(r + s)$ and $r \square s = rs$, is $(\mathbb{R}, *, \square)$ a ring?
   b) If $r * s = 2rs$ and $r \square s = rs$, is $(\mathbb{R} - \{0\}, *, \square)$ a ring?
   c) If $r * s = rs$ and $r \square s = r^s$, is $(\mathbb{R}^+, *, \square)$ a ring?

16.3 Show that the set of all real numbers of the form $a + b\sqrt{2}$, where $a, b \in \mathbb{Q}$, forms a field under ordinary addition and multiplication.

16.4 Consider $(\mathbb{Q}, *, \square)$, where $*$ is the addition given by $a * b = a + b - 1$, and $\square$ is the multiplication given by $a \square b = a + b - ab$. Is $(\mathbb{Q}, *, \square)$ a field?

16.5 (*A construction of the complex numbers.*) Let F be the set of all 2×2 matrices of the form

$$\begin{pmatrix} a & b \\ -b & a \end{pmatrix},$$

where $a, b \in \mathbb{R}$. Show that F forms a field under addition and multiplication of matrices.

Remarks. Note that if we think of

$$\begin{pmatrix} a & b \\ -b & a \end{pmatrix}$$

as representing the complex number $a + bi$, then addition and multiplication in F correspond to the usual operations on complex numbers. For example,

$$\begin{pmatrix} a & b \\ -b & a \end{pmatrix}\begin{pmatrix} c & d \\ -d & c \end{pmatrix} = \begin{pmatrix} ac - bd & ad + bc \\ -ad - bc & ac - bd \end{pmatrix},$$

which corresponds to $(ac - bd) + (ad + bc)i$. Thus this exercise shows you how to construct a field having all the desired properties of $\mathbb{C}$, starting only with $\mathbb{R}$.

16.6 Let F be a field. For $a, b \in F$, $b \neq 0$, define a/b to mean ab^{-1}. Show that:

(i) $\left(\dfrac{a}{b}\right) \cdot \left(\dfrac{c}{d}\right) = \dfrac{ac}{bd}$; (ii) $\dfrac{a}{b} + \dfrac{c}{d} = \dfrac{(ad + bc)}{bd}$

16.7 Let F be a field, let $a, b \in F$, and assume $a \neq 0$. Show that the equation

$$ax + b = 0$$

can be solved for x in F; that is, there is $x \in F$ which makes the equation true.

16.8 Prove the following for $(\mathbb{Z}_n, \oplus, \odot)$:

a) associativity for $\odot$;

b) the distributive laws.

16.9 a) Show that an element $a \in (\mathbb{Z}_n, \oplus, \odot)$ is a unit iff $(a, n) = 1$.

b) Show that every element of $\mathbb{Z}_n$ is either a unit or a zero-divisor.

c) Which elements of $\mathbb{Z}_n$ are nilpotent?

16.10 Prove parts (iv) and (v) of Theorem 16.1.

16.11 Find all units, zero-divisors, and nilpotent elements in the following rings:

a) $\mathbb{Z} \oplus \mathbb{Z}$;

b) $\mathbb{Z}_3 \oplus \mathbb{Z}_3$;

c) $\mathbb{Z}_4 \oplus \mathbb{Z}_6$.

16.12 a) Show that the trivial ring is the only ring in which 0 is not a zero-divisor.

b) Show that in any ring except the trivial ring, every nilpotent element is a zero-divisor.

16.13 a) Show that if R is a ring with unity, then the multiplicative identity element in R is unique.

b) Show that if R is a ring with unity and $a \in R$ is a unit, then the multiplicative inverse of a is unique.

16.14 An element r in a ring R is called **idempotent** if $r^2 = r$. Find all the idempotent elements in the ring of real-valued functions on $\mathbb{R}$ under addition and multiplication of functions.

16.15 (See Exercise 16.14.) Let R be a nontrivial ring with unity. Let $r \in R$ be idempotent. Show that:

a) $1 - r$ is also idempotent, and

b) either r or $1 - r$ is a zero-divisor.

16.16 Let R be a ring with unity. R is called **Boolean** if every element of R is idempotent. Show that if R is Boolean then:

a) $2r = 0$ for every $r \in R$ (in other words, $r = -r$);

b) R is commutative.

[*Hint for (a)*: Consider $(r + r)^2$.]

16.17 Let X be a set and let R be the set of all subsets of X.

a) Show that $(R, \triangle, \cap)$ is a commutative ring with unity, where $\triangle$ denotes the operation of symmetric difference.

b) Show that $(R, \triangle, \cap)$ is Boolean (see Exercise 16.16).

16.18 Let R be a ring with unity, and assume that R has no nonzero zero-divisors. Let $a, b \in R$, and assume that $ab = 1$. Show that $ba = 1$ too, and therefore a and b are units.

16.19 Let R be a ring with unity, and let $a \in R$. Assume that there is a *unique* $b \in R$ such that $ab = 1$. Show that $ba = 1$, and therefore a is a unit.

16.20 a) Let S be the set of all real-valued functions on $\mathbb{R}$. Is $(S, +, \circ)$ a ring? (Here $\circ$ denotes composition of functions.)

b) Let R be the set of all real-valued functions on $\mathbb{R}$ that are homomorphisms of the additive group $(\mathbb{R}, +)$. Is $(R, +, \circ)$ a ring?

16.21 Let G be the infinite direct product $\mathbb{Z} \times \mathbb{Z} \times \mathbb{Z} \times \mathbb{Z} \times \cdots$, where there is one copy of $\mathbb{Z}$ for each positive integer, and the operation on each copy is ordinary addition. If φ_1 and φ_2 are homomorphisms from G into itself, define $\varphi_1 + \varphi_2$ by

$$(\varphi_1 + \varphi_2)(g) = \varphi_1(g) + \varphi_2(g)$$

for all $g \in G$.

a) Show that $\varphi_1 + \varphi_2$ is a homomorphism, and that the set of all homomorphisms from G into itself forms a ring R with unity under the operations $+$ and $\circ$.

b) Show that there exist elements φ and ψ in R such that $\varphi\psi = 1$ but $\psi\varphi \neq 1$ (here 1 denotes the unity of R).

16.22 Let $(R, +, \cdot)$ be a ring, and let S be a set. Let R^S denote the set of all functions from S to R. Show that R^S forms a ring under addition and multiplication of functions.

16.23 Let R be a ring with unity, and let U denote the set of units in R. Show that U is a group under the multiplication in R.

16.24 a) Let $\mathbb{Z}[i]$ denote the set of all complex numbers of the form $a+bi$, where $a,b\in\mathbb{Z}$. Show that $\mathbb{Z}[i]$ is a commutative ring with unity under ordinary addition and multiplication of complex numbers. $\mathbb{Z}[i]$ is called the **ring of Gaussian integers**.

b) For $r=a+bi\in\mathbb{Z}[i]$, define the *norm* $N(r)$ of r by $N(r)=a^2+b^2$. Show that if $r,s\in\mathbb{Z}[i]$, then $N(rs)=N(r)N(s)$.

c) Show that $r=a+bi$ is a unit in $\mathbb{Z}[i]$ iff $N(r)=1$. Using this information, find all the units in $\mathbb{Z}[i]$.

d) (See Exercise 16.23.) The group of units of $\mathbb{Z}[i]$ is isomorphic to a familiar group. Which one?

16.25 Let R denote the ring of all real numbers of the form $a+b\sqrt{2}$, where $a,b\in\mathbb{Z}$. For $r=a+b\sqrt{2}\in R$, define $N(r)$ by $N(r)=a^2-2b^2$.

a) Show that if $r,s\in R$ then $N(rs)=N(r)N(s)$.

b) Show that r is a unit in R iff $N(r)=\pm1$.

c) Show that there are infinitely many units in R.

16.26 Give an example of a finite noncommutative ring.

16.27 Give an example of a noncommutative ring with no multiplicative identity.

16.28 Let R be an integral domain. If there exists a positive integer n such that $n\cdot1=0$, then the smallest such integer is called the **characteristic** of R. If no such n exists, then we say that R has characteristic 0.

a) Show that if R has characteristic n, then $n\cdot r=0$ for every $r\in R$.

b) Show that if R has characteristic $n>0$, then n is a prime number.

c) For each prime number p, give an example of a field of characteristic p. Give an example of a field of characteristic 0.

16.29 Must every ring with a prime number of elements be commutative? Either prove that it must, or give a counterexample.

16.30 Let R be a finite nontrivial ring with no nonzero zero-divisors. Show that R is a division ring.

SUBRINGS, IDEALS, AND QUOTIENT RINGS

In this section we shall develop analogues, for rings, of some of the concepts we encountered in dealing with groups. As for groups, the purpose of doing this is to develop ways of talking about the internal structure of a given ring and the relationships between different rings.

We begin with the analogue of "subgroup."

DEFINITION Let $(R, +, \cdot)$ be a ring. A subset S of R is called a **subring** of R if the elements of S form a ring under $+$ and $\cdot$.

In particular, the definition requires that $(S, +)$ be a subgroup of $(R, +)$. Thus if S is a subring of R, then we know that the additive identity element 0 of R is in S, and that S is closed under addition and under additive inverses. The relationship between R and S with respect to multiplication need not be so clean. For example, if R has a multiplicative identity 1, then 1 need not be in S, and it is even possible that S may have an identity element different from that of R.

Examples

1. Let $(\mathbb{Z}, +, \cdot)$ be the integers under ordinary addition and multiplication, and let $2\mathbb{Z}$ be the set of even integers. Then it is easy to see that $2\mathbb{Z}$ is a subring of $\mathbb{Z}$. Although $\mathbb{Z}$ is a ring with unity, $2\mathbb{Z}$ has no unity.

2. Consider the ring $\mathbb{R} \oplus \mathbb{R}$. Let S denote the set of all pairs of the form $(r, 0)$, where $r \in \mathbb{R}$. Then S is a subring of $\mathbb{R} \oplus \mathbb{R}$, and $\mathbb{R} \oplus \mathbb{R}$ has unity $(1, 1)$, which is not in S. In this case, S has its own unity, namely $(1,0)$. Note that $(1,0)$ is *not* a unity for $\mathbb{R} \oplus \mathbb{R}$, but only for S.

As for subgroups, the definition of "subring" can be recast in more compact form.

THEOREM 17.1 Let $(R, +, \cdot)$ be a ring, and let S be a subset of R. Then S is a subring of R iff the following two conditions are satisfied:

i) $(S, +)$ is a subgroup of $(R, +)$; and

ii) S is closed under multiplication, that is, if $r_1, r_2 \in S$ then $r_1 r_2 \in S$.

PROOF. It is clear that if S is a subring of R, then (i) and (ii) hold. Conversely, assume that (i) and (ii) hold for S. Then $+$ and $\cdot$ are both binary operations on S, and $(S, +)$ is an abelian group, so we need only check that $\cdot$ is associative on S and that the distributive laws hold in $(S, +, \cdot)$. But associativity and distributivity hold for all elements of R, hence for those of S. $\square$

Condition (i) can be reduced further by using your favorite subgroup criteria. For example, by using the result of Exercise 5.24 we obtain the following.

COROLLARY 17.2 Let $(R, +, \cdot)$ be a ring and let S be a nonempty subset of R. Then S is a subring of R iff the following two conditions hold:

i) for every $r_1, r_2 \in S$, we have $r_1 - r_2 \in S$;

ii) for every $r_1, r_2 \in S$, we have $r_1 r_2 \in S$.

Examples (continued)

3. Let's find all the subrings of $(\mathbb{Z}, +, \cdot)$. We know that any subring must be a subgroup of $(\mathbb{Z}, +)$ and hence must be additively a cyclic subgroup of the form $m\mathbb{Z}$, for some m. We have only to figure out which of these subgroups constitute subrings. The extra requirement that $m\mathbb{Z}$ must satisfy to be a subring is closure under multiplication. But clearly if i, j are integers, then $(im)(jm) = (ijm)m \in m\mathbb{Z}$. Thus every subgroup of $(\mathbb{Z}, +)$ is a subring of $(\mathbb{Z}, +, \cdot)$.

Similar reasoning shows that every subgroup of $(\mathbb{Z}_n, \oplus)$ is a subring of $(\mathbb{Z}_n, \oplus, \odot)$.

4. Let R be the set of all real-valued functions defined on $\mathbb{R}$ under pointwise addition and multiplication of functions. Let S be the subset of R consisting of all the continuous functions. Then S is a subring of R, for if f, g are continuous functions so is $f - g$ and so is fg. (We have used Corollary 17.2.)

5. Let R be as in the previous example and let

$$S = \{f \in R \mid f(0) = 0\},$$

so that S consists of those functions that take the value 0 at $x = 0$. Then S is a subring of R, for if $f, g \in S$ then $f(0) = 0$ and $g(0) = 0$, so

$$(f - g)(0) = f(0) - g(0) = 0,$$

which shows that $f - g \in S$, and

$$fg(0) = f(0)g(0) = 0,$$

which shows that $fg \in S$.

Observe that if $T = \{f \in R \mid f(0) = 1\}$, then T is not a subring of R, because if $f(0) = g(0) = 1$ then $(f - g)(0) = 1 - 1 = 0$.

6. Let $M_2(\mathbb{R})$ be the ring of all 2×2 matrices with real entries. Let S consist of all matrices of the form $\begin{pmatrix} a & b \\ 0 & d \end{pmatrix}$. Then S is a subring of $M_2(\mathbb{R})$, for if $\begin{pmatrix} a & b \\ 0 & d \end{pmatrix}$ and $\begin{pmatrix} e & f \\ 0 & h \end{pmatrix}$ are in S, then so is $\begin{pmatrix} a & b \\ 0 & d \end{pmatrix} - \begin{pmatrix} e & f \\ 0 & h \end{pmatrix}$, and so is $\begin{pmatrix} a & b \\ 0 & d \end{pmatrix}\begin{pmatrix} e & f \\ 0 & h \end{pmatrix}$.

On the other hand, if T consists of all matrices of the form $\begin{pmatrix} a & b \\ c & 0 \end{pmatrix}$, then T is not closed under multiplication, so T is not a subring of $M_2(\mathbb{R})$.

7. Let $M_2(\mathbb{C})$ be the ring of all 2×2 matrices with complex entries. Let

$$I = \begin{pmatrix} 1 & 0 \\ 0 & 1 \end{pmatrix}, \quad J = \begin{pmatrix} i & 0 \\ 0 & -i \end{pmatrix}, \quad K = \begin{pmatrix} 0 & 1 \\ -1 & 0 \end{pmatrix}, \quad \text{and} \quad L = JK = \begin{pmatrix} 0 & i \\ i & 0 \end{pmatrix}.$$

Let $\mathbb{H}$ be the following subset of $M_2(\mathbb{C})$:

$$\mathbb{H} = \{aI + bJ + cK + dL \mid a, b, c, d \in \mathbb{R}\}.$$

(*Note*: The product of a constant and a matrix is obtained by multiplying each entry of the matrix by the constant.) $\mathbb{H}$ is a subring of $M_2(\mathbb{C})$, for it is clear that $\mathbb{H}$ is an additive subgroup, and closure of $\mathbb{H}$ under multiplication follows from distributivity in $M_2(\mathbb{C})$ and the fact that $\{\pm I, \pm J, \pm K, \pm L\}$ is closed under multiplication. $[\{\pm I, \pm J, \pm K, \pm L\}$, under multiplication, is Q_8.]

$\mathbb{H}$ is called the **ring of quaternions**, and "$\mathbb{H}$" is for Hamilton, the man who discovered this ring. $\mathbb{H}$ is a noncommutative ring, for the same reason that Q_8 is a nonabelian group: $JK \neq KJ$. $\mathbb{H}$ has an identity $I \neq \begin{pmatrix} 0 & 0 \\ 0 & 0 \end{pmatrix}$, and it is easy to see that $\mathbb{H}$ is in fact a division ring. For if

$$aI + bJ + cK + dL = \begin{pmatrix} a + bi & c + di \\ -c + di & a - bi \end{pmatrix} \in \mathbb{H}$$

and at least one of a, b, c, d is not zero, we have

$$\begin{pmatrix} a+bi & c+di \\ -c+di & a-bi \end{pmatrix}^{-1} = \frac{1}{a^2+b^2+c^2+d^2} \begin{pmatrix} a-bi & -c-di \\ c-di & a+bi \end{pmatrix}$$

$$= \frac{1}{a^2+b^2+c^2+d^2}(aI - bJ - cK - dL) \in \mathbb{H}.$$

$\mathbb{H}$ is the example of a noncommutative division ring that we promised you in Section 16.

Hamilton discovered the quaternions in 1843, after he had spent ten or fifteen years seeking a generalization of $\mathbb{C}$ that could be used in connection with geometric and physical problems in 3-space. One reason why it took him so long was that he started out looking for a commutative generalization; coming up with a noncommutative one was, at the time, a revolutionary step.

The definition of $\mathbb{H}$ via $M_2(\mathbb{C})$ was not possible until 1858, when Cayley introduced matrices. Hamilton thought of $\mathbb{H}$ while taking a stroll on the evening of October 16, 1843; it occurred to him as a set of elements of the form

$$a + bi + cj + dk,$$

where $a, b, c, d \in \mathbb{R}$ and

$$i^2 = j^2 = k^2 = ijk = -1.$$

Although he made other distinguished contributions to science, Hamilton considered the discovery of the quaternions to be the crowning achievement of his life. He spent twenty years studying them, and wrote several huge volumes about them.

In dealing with groups, we found that some subgroups were better than others. For example, in attempting to construct the quotient group modulo a subgroup H, we saw that it was crucial for H to be normal. We encounter a similar situation when we try to construct quotient rings.

Let $(R, +, \cdot)$ be a ring and let S be a subring of R. We know that $(S, +)$ is a subgroup of $(R, +)$, and in fact there is no problem about $(S, +)$ being a *normal* subgroup of $(R, +)$, because $(R, +)$ is abelian. Thus we already know how to construct a quotient group $(R/S, +)$, the elements of which are the cosets of S in R, with addition defined by

$$(S+a)+(S+b) = S+(a+b).$$

We would like to endow this quotient group with a multiplication, arising naturally from the given multiplication in R, in such a way that the quotient becomes a ring. As usual with these things, there is only one reasonable attempt; we want to define

$$(S + a)(S + b) = S + ab.$$

What we have to check is that this is a well-defined operation, that is, if

$$S + a = S + a' \text{ and } S + b = S + b',$$

then

$$S + ab = S + a'b',$$

so that the product doesn't depend on which representatives we use to define it.

So we assume $a - a' \in S$ and $b - b' \in S$, and we wish to show that $ab - a'b' \in S$. It is clear that if this is to work for all possible choices of a, a', b, b', then S has to be rather special. For instance, if we take $a \in S$, $a' = 0$, b arbitrary, and $b' = b$, then

$$a - a' \in S \quad \text{and} \quad b - b' \in S,$$

so we want

$$ab - 0b \in S, \quad \text{that is,} \quad ab \in S.$$

In other words, we require that if $a \in S$ and b is *any* element of R, then $ab \in S$. Similar reasoning shows that we also require that if $b \in S$ and $a \in R$ is arbitrary, then $ab \in S$.

Now we claim that these two conditions are enough to make our multiplication work out. For suppose S satisfies both conditions. Assume $a - a' \in S$ and $b - b' \in S$. Then

$$ab - a'b' = (a - a')b + a'(b - b').$$

By the conditions on S, $(a - a')b \in S$ and $a'(b - b') \in S$, so

$$(a - a')b + a'(b - b') \in S$$

since S is an additive subgroup. Thus $ab - a'b' \in S$, and our multiplication is well defined.

Subrings that have the special properties required to make multiplication of the additive cosets well defined are called *ideals*.

DEFINITION A subring S of a ring R is called an **ideal** of R if for every $s \in S$ and $r \in R$ we have $rs \in S$ and $sr \in S$.

We have seen that if S is a subring of R, then the natural attempt at introducing a multiplication on R/S will succeed iff S is an ideal. It is now easy to complete the proof of

THEOREM 17.3 Let $(R, +, \cdot)$ be a ring, and let S be an ideal of R. Then the set R/S of cosets of the additive subgroup $(S, +)$ is a ring under the operations

$$(S + a) + (S + b) = S + (a + b),$$
$$(S + a)(S + b) = S + ab.$$

PROOF. We know that R/S is a group under the indicated addition, and this group is abelian because $(R, +)$ is. We have seen above that the indicated multiplication yields a binary operation on R/S, so all that we have to check is that this multiplication is associative and that the distributive laws hold. Associativity requires that

$$[(S + a)(S + b)](S + c) = (S + a)[(S + b)(S + c)]$$

for all $a, b, c \in R$. This amounts to

$$(S + ab)(S + c) = (S + a)(S + bc), \quad \text{that is,}$$
$$S + (ab)c = S + a(bc),$$

which is true by associativity of multiplication in R. Similarly, R/S inherits distributivity from R; we leave the details of this to the reader. $\square$

R/S is called the **quotient ring** (or **factor ring**) of R by S.

There is a certain redundancy in our definition of "ideal," in that the condition that $rs \in S$ and $sr \in S$ for every $r \in R$ and $s \in S$ already implies part of the condition that S be a subring. The next result provides a neater characterization of ideals.

THEOREM 17.4 Let R be a ring and let S be a subset of R. Then S is an ideal of R iff the following two conditions hold:

i) S in an additive subgroup of R (equivalently, S is nonempty and closed under subtraction);

ii) For every $r \in R$ and $s \in S$, we have $rs \in S$ and $sr \in S$.

PROOF. Exercise.

We will usually denote an ideal by I, rather than S, from now on.

Examples (continued)

8. As we have seen, the subrings of $(\mathbb{Z}, +, \cdot)$ are precisely the additive subgroups $m\mathbb{Z}$. In this case [likewise for $(\mathbb{Z}_n, \oplus, \odot)$], every subring is also an ideal, because if we multiply an element of $m\mathbb{Z}$ by any integer we get an element of $m\mathbb{Z}$.

9. $\mathbb{Z}$ is a subring of $(\mathbb{Q}, +, \cdot)$, but not an ideal. For example, $1 \in \mathbb{Z}$ and $\frac{1}{2} \in \mathbb{Q}$, but $1 \cdot \frac{1}{2} \notin \mathbb{Z}$.

Actually, there is something more general going on in this example. If R is a ring with unity 1, then the only ideal of R that contains 1 is R itself. (Prove it!)

Incidentally, R will always be an ideal of R, no matter what ring R is. We will call R the **improper** ideal; all other ideals are called **proper**. The ideal $\{0\}$ is called **trivial**.

10. Refer back to Examples 4 and 5, which present two examples of subrings of the ring R of all real-valued functions on $\mathbb{R}$. The subring S in Example 4 is not an ideal, because if $f \in S$ and $g \in R$ then fg need not be in S, since it need not be continuous. (Can you give an example?) The subring S in Example 5 is an ideal, however, because if $f(0) = 0$ and g is any element of R, then

$$fg(0) = f(0)g(0) = 0 \cdot g(0) = 0,$$

so $fg \in S$. Likewise $gf \in S$.

11. Let I be the subring of $\mathbb{R} \oplus \mathbb{R}$ consisting of all pairs of the form $(r, 0)$. Then I is an ideal, because $(r, 0)(a, b) = (ra, 0) \in I$ for any $r, a, b \in \mathbb{R}$, and likewise $(a, b)(r, 0) \in I$. Note that I has a unity, and yet I is a *proper* ideal of R. Why does this not contradict our observation in Example 9 about ideals that contain the unity?

12. Let R be a commutative ring with unity, and let $a \in R$. Let

$$aR = \{ar \mid r \in R\},$$

so that aR is the set of all multiples of a in R. Then aR is an ideal of R. First of all, if $ar_1, ar_2 \in aR$ then $ar_1 - ar_2 = a(r_1 - r_2) \in aR$, so aR is an additive subgroup of R. Secondly, if $ar \in aR$ and $t \in R$, then

$$t(ar) = (ar)t = a(rt) \in aR,$$

since R is commutative, and this verifies that aR is an ideal.

We call aR the **principal ideal generated by** a. Note that $a \in aR$, since $a = a \cdot 1$. Also observe that $aR = Ra = \{ra \mid r \in R\}$.

13. Let $R = (\mathbb{Z}, +, \cdot)$ and let I be the ideal $n\mathbb{Z}$ for some positive integer n. Then $(R/I, +, \cdot)$ has n elements, namely, $I + 0, I + 1, I + 2, \ldots, I + (n - 1)$. We saw in Section 13 that $(R/I, +)$ is isomorphic, as a group, to $(\mathbb{Z}_n, \oplus)$. In Section 18 we shall introduce a notion of isomorphism for rings, and it will turn out that $(\mathbb{Z}/n\mathbb{Z}, +, \cdot)$ is isomorphic, as a ring, to $(\mathbb{Z}_n, \oplus, \odot)$.

Suppose now that R is a ring and I is an ideal of R. Then R/I is a ring, and we can sensibly (and profitably) ask which properties in R translate into familiar properties for the elements of R/I. For example, if $a \in R$, when is $I + a$ a zero-divisor in R/I?

$I + a$ is a zero-divisor iff there is some $I + b \neq I + 0$ such that either $(I+a)(I+b) = I+0$ or $(I+b)(I+a) = I+0$. This boils down to there being some $b \notin I$ such that either $ab \in I$ or $ba \in I$. In particular, $I + a$ is a nontrivial (i.e., nonzero) zero-divisor in R/I iff $a \notin I$ and there is some $b \notin I$ such that $ab \in I$ or $ba \in I$.

From this it is clear what conditions we need on I to rule out nontrivial zero-divisors in R/I.

DEFINITION Let R be a ring, I an ideal in R. Then I is **prime** if whenever $a, b \in R$ and $ab \in I$, then at least one of a or b is in I.

Example Let p be a prime in $\mathbb{Z}$. Then $p\mathbb{Z}$ is a prime ideal, because if ab is divisible by p, then one of a or b must be divisible by p.

THEOREM 17.5 R/I has no nontrivial zero-divisors iff I is prime.

The proof is immediate from the above discussion. Specializing to the case where R is a commutative ring with unity, we get:

COROLLARY 17.6 Let R be a commutative ring with unity, I an ideal in R. Then R/I is an integral domain iff I is a proper prime ideal.

PROOF. R/I is a commutative ring with unity (see Exercise 17.14). Thus R/I is an integral domain iff it is nontrivial and has no nontrivial zero-divisors, that is, iff I is proper and prime. $\square$

When is R/I a field?

DEFINITION. An ideal I of R is called **maximal** if I is proper and there is no proper ideal $J \supsetneq I$.

Thus I is maximal iff it is proper and cannot be extended to a larger proper ideal.

THEOREM 17.7 Let R be a commutative ring with unity. If I is an ideal in R, then R/I is a field iff I is maximal.

PROOF. R/I is a commutative ring with unity. Thus it is a field iff it is nontrivial and each of its nonzero elements is a unit.

Now R/I is nontrivial iff I is proper. And every nonzero element of R/I is a unit $\Leftrightarrow$ for every $a \in R - I$, there is $b \in R$ such that $(I+a)(I+b) = I+1$, in other words,

$$\text{for every } a \notin I, \text{ there is } b \text{ such that } ab - 1 \in I. \qquad \text{[17.1]}$$

Thus R/I is a field iff I is proper and [17.1] holds. To conclude the proof, we will show that a proper ideal I is maximal iff [17.1] holds.

First suppose that I is maximal, and take $a \notin I$. Then the set

$$J = \{ ar + x \mid r \in R \text{ and } x \in I \}$$

is an ideal that properly includes I, hence is not proper. Therefore, $ar_0 + x_0 = 1$ for some r_0 and x_0, and this yields

$$ar_0 - 1 = -x_0 \in I,$$

which means that [17.1] holds. Conversely, if [17.1] holds, then let I' be an ideal such that $I' \supsetneq I$, with the aim of showing that $I' = R$. Take $a \in I' - I$, and take b such that $ab - 1 = y \in I$. Then $a \in I'$ and $y \in I'$, so $ab - y \in I'$. Thus $1 \in I'$, so $I' = R$, and I is maximal. $\square$

COROLLARY 17.8 Let R be a commutative ring with unity. Then every maximal ideal of R is prime.

PROOF. If I is maximal, then R/I is a field, hence an integral domain. Thus I is prime by Corollary 17.6. $\square$

This last result can fail when we try to weaken the assumptions. For instance, let R be the ring on $(\mathbb{Z}_2, \oplus)$ with trivial multiplication. Then R is a commutative ring *without* unity, and $\{0\}$ is a maximal ideal which is not prime.

On the other hand, prime ideals need not be maximal, even if R is a commutative ring with unity. According to our definitions, in fact, any ring is a prime ideal of itself which is not maximal. If this seems like cheating, note that $\{0\}$ is a prime ideal in $\mathbb{Z}$ which is not maximal. And for an example of a *nontrivial* proper prime ideal which is not maximal, take the subset I of $\mathbb{Z} \oplus \mathbb{Z}$ consisting of all pairs $(a, 0)$. Note that

$$\{ (a, 2b) \mid a, b \in \mathbb{Z} \}$$

is a proper ideal which is strictly larger than I.

Some conditions under which proper prime ideals *are* maximal are indicated in Exercises 17.29–17.31.

Here are a few more examples of maximal ideals.

Examples

1. We have observed that every ideal in $\mathbb{Z}$ has the form $n\mathbb{Z}$ for some n, and that if n is a prime p, then $p\mathbb{Z}$ is a prime ideal. If $n \geqslant 2$ is not prime, then $n\mathbb{Z}$ is clearly not a prime ideal, so the only other prime ideals in $\mathbb{Z}$ are $\{0\}$ and $\mathbb{Z}$.

What are the maximal ideals? Since $\mathbb{Z}$ is a commutative ring with unity, any maximal ideal must be prime; and since $\{0\}$ and $\mathbb{Z}$ are obviously not maximal, the only possible maximal ideals are of the form $p\mathbb{Z}$. In fact every $p\mathbb{Z}$ is maximal. For suppose $J \supsetneq p\mathbb{Z}$ and J is an ideal. If $x \in J - p\mathbb{Z}$ then $(x,p) = 1$, so there are a, b such that $ax + bp = 1$. Since $ax \in J$ and $bp \in J$, $1 \in J$, so $J = \mathbb{Z}$.

Thus for $\mathbb{Z}$ the maximal ideals are the nontrivial proper prime ideals.

2. The ideal $I = \{(a, 2b) | a, b \in \mathbb{Z}\}$ is maximal in $\mathbb{Z} \oplus \mathbb{Z}$. For if J is an ideal properly including I, we have $(m, 2n+1) \in J$ for some $m, n \in \mathbb{Z}$. Since $(m, 2n) \in I$, subtracting shows that $(0, 1) \in J$. Also, $(1, 0) \in I \subseteq J$. Adding, we see that $(1, 1) \in J$, so $J = \mathbb{Z} \oplus \mathbb{Z}$.

3. Let us return to Example 5 above. R is the ring of all real-valued functions on $\mathbb{R}$, under pointwise addition and multiplication, and I is the ideal consisting of all f such that $f(0) = 0$. We want to show that I is maximal. Suppose J is an ideal that is strictly larger than I, and take $g \in J - I$. Then $g(0) \neq 0$, so there is $h \in R$ such that

$$h(0) = \frac{1}{g(0)}.$$

We have $g(0)h(0) = 1$, so if f_1 denotes the multiplicative identity in R, the value of the function $gh - f_1$ at 0 is 0. Hence

$$gh - f_1 \in I \subseteq J,$$

so since $gh \in J$, $f_1 \in J$. Thus $J = R$, and I is maximal.

The same argument demonstrates that for any $r \in \mathbb{R}$, $I_r = \{f \in R | f(r) = 0\}$ is a maximal ideal in R.

4. Let S be the subring of $M_2(\mathbb{R})$ introduced in Example 6 above, that is, $S = \left\{ \begin{pmatrix} a & b \\ 0 & d \end{pmatrix} | a, b, d \in \mathbb{R} \right\}$. We contend that $I = \left\{ \begin{pmatrix} a & b \\ 0 & 0 \end{pmatrix} | a, b \in \mathbb{R} \right\}$ is a maximal ideal of S. First of all, it is clear that I is closed under subtraction in S, and if $\begin{pmatrix} e & f \\ 0 & h \end{pmatrix} \in S$ and $\begin{pmatrix} a & b \\ 0 & 0 \end{pmatrix} \in I$, we have

$$\begin{pmatrix} e & f \\ 0 & h \end{pmatrix}\begin{pmatrix} a & b \\ 0 & 0 \end{pmatrix} = \begin{pmatrix} ea & eb \\ 0 & 0 \end{pmatrix} \in I \quad \text{and}$$

$$\begin{pmatrix} a & b \\ 0 & 0 \end{pmatrix}\begin{pmatrix} e & f \\ 0 & h \end{pmatrix} = \begin{pmatrix} ae & af + bh \\ 0 & 0 \end{pmatrix} \in I,$$

so I is an ideal. To see that I is maximal, let J be an ideal of S that properly includes I. Then we have $\begin{pmatrix} a & b \\ 0 & d \end{pmatrix} \in J$ for some a, b, d, with $d \neq 0$, and therefore

$$\begin{pmatrix} 0 & 0 \\ 0 & 1/d \end{pmatrix}\begin{pmatrix} a & b \\ 0 & d \end{pmatrix} = \begin{pmatrix} 0 & 0 \\ 0 & 1 \end{pmatrix} \in J.$$

Since $\begin{pmatrix} 1 & 0 \\ 0 & 0 \end{pmatrix} \in I$, we get

$$\begin{pmatrix} 1 & 0 \\ 0 & 0 \end{pmatrix} + \begin{pmatrix} 0 & 0 \\ 0 & 1 \end{pmatrix} = \begin{pmatrix} 1 & 0 \\ 0 & 1 \end{pmatrix} \in J,$$

so $J = S$, as desired.

EXERCISES

17.1 Which of the following subsets of $M_2(\mathbb{R})$ are subrings?

   a) $S =$ all matrices of the form $\begin{pmatrix} 0 & b \\ c & d \end{pmatrix}$

   b) $S =$ all matrices of the form $\begin{pmatrix} a & 0 \\ c & d \end{pmatrix}$

   c) $S = GL(2, \mathbb{R})$

   d) $S =$ all matrices of the form $\begin{pmatrix} a & b \\ b & a \end{pmatrix}$.

17.2 Let R be the ring of real-valued functions on the real line, under pointwise operations. Which of the following subsets S of R are subrings? Which are ideals?

   a) $S = \{ f \in R \mid f(1) = 0 \}$

   b) $S = \{ f \in R \mid f(1) = 0 \text{ or } f(2) = 0 \}$

   c) $S = \{ f \in R \mid f(3) = f(4) \}$.

17.3 Let R be the ring with trivial multiplication on some abelian group G. What are the ideals of R?

17.4 Let S, T be subrings of R. Under what conditions is $S \cup T$ a subring of R?

17.5 Let R be a finite ring with, say, n elements. Let S be a subring of R that has m elements. Show that m divides n.

17.6 Find a maximal ideal in

   a) $\mathbb{Z}_6$;   b) $\mathbb{Z}_{12}$;   c) $\mathbb{Z}_{18}$.

17.7 Find all the maximal ideals in $\mathbb{Z}_n$.

17.8 Let $R = 2\mathbb{Z}$ be the ring of even integers, under ordinary addition and multiplication, and consider the subset $S = 4\mathbb{Z}$. Show that S is an ideal of R. Is S maximal? Is it prime?

17.9 Let $R = \{ q \in \mathbb{Q} \mid q = a/b, \ a, b \in \mathbb{Z} \text{ and } b \text{ is odd} \}$. Show that R has a unique maximal ideal.

17.10 Let X be a nonempty set and let R be the ring $(P(X), \triangle, \cap)$.

   a) Show that if $Y \subsetneq X$, then $P(Y)$ is an ideal in R and has a unity different from that of R.

   b) Find a maximal ideal in R.

17.11 Prove Theorem 17.4.

17.12 Let R be a ring with unity. Show that a nonempty subset S of R is an ideal iff the following two conditions are satisfied:

i) $s_1 + s_2 \in S$ for every $s_1, s_2 \in S$;

ii) rs and sr are in S for every $s \in S$, $r \in R$.

17.13 Show that if I is an ideal of R, then the distributive laws hold in R/I.

17.14 Let R be a ring and I an ideal of R. Show that

a) if R is commutative, so is R/I;

b) if R has a unity, so does R/I.

17.15 Show that if S is a subring of R and I is an ideal such that $I \subseteq S$, then S/I is a subring of R/I. Show that if S is an ideal of R, then S/I is an ideal of R/I.

17.16 Show that $M_2(\mathbb{R})$ has no ideals other than the trivial ideal and the improper ideal.

17.17 Let S be the ring of all matrices of the form $\begin{pmatrix} a & b \\ 0 & d \end{pmatrix}$, with $a, b, d \in \mathbb{R}$. Find a maximal ideal of S other than the one found in the text, and show that every proper ideal of S is contained in one of these two maximal ideals.

17.18 Let $\mathbb{Z}[i]$ be the ring of Gaussian integers (see Exercise 16.24). Let I be the principal ideal generated by the element $2 + 2i$. How many elements are there in $\mathbb{Z}[i]/I$?

17.19 Let R be a commutative ring with unity $1 \neq 0$.

a) Show that R is a domain iff $\{0\}$ is a prime ideal in R.

b) Show that R is a field iff $\{0\}$ is a maximal ideal in R.

17.20 Let R be a commutative ring with unity, and let $a \in R$. Show that $aR = R$ iff a is a unit.

17.21 a) Let R be a ring. Define the **center** of R to consist of all $r \in R$ such that $rx = xr$ for every $x \in R$. Show that the center of R is a subring of R.

b) Must the center of a ring be an ideal?

17.22 a) Let R be a commutative ring and X a subset of R. Define
$$\text{Ann}(X) = \{r \in R \mid rx = 0 \text{ for every } x \in X\}.$$
$\text{Ann}(X)$ is called the **annihilator** of X. Show that $\text{Ann}(X)$ is an ideal.

b) In $(\mathbb{Z}_{12}, \oplus, \odot)$, find $\text{Ann}(\{2\})$.

c) If R were not commutative, then the set $\text{Ann}(X)$ defined above would be called the *left* annihilator of X. Show that in this case, if X is itself an ideal, then $\text{Ann}(X)$ is still an ideal.

17.23 a) Let $R_1, R_2, \dots, R_n$ be rings with unity. Show that every ideal of $R_1 \oplus R_2 \oplus \cdots \oplus R_n$ is of the form $I_1 \oplus I_2 \oplus \cdots \oplus I_n$, where I_i is an ideal of R_i for each i.

b) Show that this result may fail if the R_i's are not rings with unity.

17.24 Let $R_1, R_2, \ldots, R_n$ be rings with unity. What do the maximal ideals of $R_1 \oplus R_2 \oplus \cdots \oplus R_n$ look like?

17.25 a) Let R be a ring and let I and J be ideals of R. Show that $I \cap J$ is an ideal of R.

b) Suppose that I and J are prime. Must $I \cap J$ be prime?

17.26 a) Give an example of a ring R, an ideal I of R, and an ideal J of I such that J is not an ideal of R.

b) Show that J must be an ideal of R if it is a prime ideal of I.

17.27 Let R be a commutative ring.

a) Show that the set of nilpotent elements in R forms an ideal.

b) Show that the quotient of R by this ideal has no nonzero nilpotent elements.

17.28 Let R be a commutative ring, and let I be an ideal of R. Define the **radical**, $\sqrt{I}$, of I to consist of all $r \in R$ such that some power of r is in I.

a) Show that $\sqrt{I}$ is an ideal of R.

b) I is called **semiprime** if $I = \sqrt{I}$. Show that I is semiprime iff R/I has no nontrivial nilpotent elements.

17.29 Let R be a finite commutative ring with unity. Show that every proper prime ideal of R is maximal.

17.30 Let R be an integral domain. We call R a **principal ideal domain** (PID) if every ideal of R is of the form aR for some $a \in R$. Show that in a PID every nontrivial proper prime ideal is maximal.

17.31 Let R be a Boolean ring (see Exercise 16.16). Show that every proper prime ideal of R is maximal.

17.32 Prove that every nontrivial finite subring of a division ring is a division ring.

17.33 a) Let I and J be ideals of R. Define their **sum** $I + J$ by

$$I + J = \{x + y \mid x \in I, y \in J\}.$$

Show that $I + J$ is an ideal.

b) Find $6\mathbb{Z} + 14\mathbb{Z}$ in $(\mathbb{Z}, +, \cdot)$.

17.34 Let I and J be ideals of R. Define their **product** IJ to be the set of all finite sums of the form $x_1 y_1 + \cdots + x_n y_n$, where each $x_i \in I$ and each $y_i \in J$. Show that IJ is an ideal, and that $IJ \subseteq I \cap J$.

17.35 Let I, J, K be ideals of R. Assume that $IJ \subseteq K$ and K is prime. Show that at least one of I or J is contained in K.

17.36 Let R be a commutative ring with unity, and let I and J be ideals of R such that $I + J = R$. Show that $IJ = I \cap J$.

17.37 Let I, J, K be ideals of R. Show that $I(J + K) = IJ + IK$.

RING HOMOMORPHISMS

We have seen that group homomorphisms enable us to relate different groups to each other. Ring homomorphisms do the same thing for rings.

In the context of rings, a homomorphism must be "sensible" with respect to both operations:

DEFINITION Let R and S be rings, and let $\varphi: R \to S$ be a function. Then φ is called a **(ring) homomorphism** if for every $a, b \in R$ we have

i) $\varphi(a + b) = \varphi(a) + \varphi(b)$ and

ii) $\varphi(ab) = \varphi(a)\varphi(b)$.

Thus a ring homomorphism is in particular a group homomorphism from $(R, +)$ into $(S, +)$. As such, it has many familiar properties. For instance, $\varphi(0_R) = 0_S$, and $\varphi(na) = n\varphi(a)$ for every $a \in R$ and $n \in \mathbb{Z}$.

A ring homomorphism φ must also preserve products, but in general this doesn't have as many consequences as the corresponding fact for sums, because $(R - \{0\}, \cdot)$ and $(S - \{0\}, \cdot)$ need not be groups. It is entirely possible, for instance, that R and S are both rings with unity, and yet $\varphi(1_R) \neq 1_S$. Because of this, it is possible that $u \in R$ is a unit but $\varphi(u) \in S$ is not.

Examples Let $\varphi: \mathbb{R} \to \mathbb{R}$ be given by $\varphi(r) = 0$ for all $r \in \mathbb{R}$. Then φ is a ring homomorphism, $\varphi(1_{\mathbb{R}}) \neq 1_{\mathbb{R}}$, and $\varphi(r)$ is never a unit, although every $r \neq 0$ is a unit in $\mathbb{R}$.

For a slightly less trivial example, let

$$\varphi: \mathbb{R} \to \mathbb{R} \oplus \mathbb{R}$$

be given by $\varphi(r) = (r, 0)$ for every $r \in \mathbb{R}$. Clearly φ is a ring homomorphism, but $\varphi(1_{\mathbb{R}}) = (1, 0)$, which is not the multiplicative identity in $\mathbb{R} \oplus \mathbb{R}$. Again φ maps all the units in $\mathbb{R}$ to nonunits.

Many times people who are working solely in the context of rings with unity tighten up their definition of homomorphism so as to insist that $\varphi(1_R) = 1_S$. This is nice in that it avoids maps like the φ's in our examples, but in our general context our definition is probably better, and we will stick to it.

We record some basic properties of ring homomorphisms for reference.

THEOREM 18.1 Let $\varphi: R \to S$ be a ring homomorphism. Then:

  i) $\varphi(0_R) = 0_S$;

 ii) $\varphi(na) = n\varphi(a)$, for every $a \in R$ and $n \in \mathbb{Z}$;

iii) $\varphi(a^n) = [\varphi(a)]^n$ for every $a \in R$ and every positive integer n.

iv) If R and S are rings with unity and $\varphi(1_R) = 1_S$, then for every unit $u \in R$, $\varphi(u)$ is a unit in S and $\varphi(u^{-1}) = [\varphi(u)]^{-1}$. More generally, $\varphi(u^n) = [\varphi(u)]^n$ for every integer n. (Here, for $n < 0$, u^n is defined to be $(u^{-1})^{|n|}$.)

The proofs of parts (iii) and (iv) are left as exercises. In light of (iv) it is of some interest to have conditions under which $\varphi(1_R)$ will equal 1_S.

THEOREM 18.2 Let R and S be rings with unity, and let $\varphi: R \to S$ be a ring homomorphism. Then:

  i) if φ is onto, $\varphi(1_R) = 1_S$;

 ii) If S is a division ring and $\varphi(1_R) \neq 0_S$, then $\varphi(1_R) = 1_S$.

iii) if S is an integral domain and $\varphi(1_R) \neq 0_S$, then $\varphi(1_R) = 1_S$.

PROOF. i) If we knew that $\varphi(1_R) \cdot s = s \cdot \varphi(1_R) = s$ for every $s \in S$, then we could conclude that $\varphi(1_R) = 1_S$, by the uniqueness of the multiplicative identity. Since φ is onto, any s must be $\varphi(r)$ for some $r \in R$, and we know that

$$1_R \cdot r = r \cdot 1_R = r.$$

Thus

$$\varphi(1_R) \cdot \varphi(r) = \varphi(r) \cdot \varphi(1_R) = \varphi(r),$$

which gives us $\varphi(1_R) \cdot s = s \cdot \varphi(1_R) = s$, as desired.

ii) We know that $1_R = 1_R \cdot 1_R$, so

$$\varphi(1_R) = \varphi(1_R) \cdot \varphi(1_R).$$

Since S is a division ring and $\varphi(1_R) \neq 0_S$, this yields $\varphi(1_R) = 1_S$, because we can multiply both sides of the equation by $[\varphi(1_R)]^{-1}$.

iii) Exercise. $\square$

As in the case of groups, we give special names to ring homomorphisms φ that satisfy extra conditions. Thus, φ is a **monomorphism** (or an **embedding**) if it is one-to-one; φ is an **epimorphism** if it is onto; and φ is an **isomorphism** if it is both one-to-one and onto. Two rings R and S are said to be **isomorphic** if there exists an isomorphism from R onto S. In this case we write $R \cong S$.

Examples

1. Let $\varphi:(\mathbb{Z}_8, \oplus, \odot) \to (\mathbb{Z}_4, \oplus, \odot)$ be given by $\varphi(x) =$ the remainder of x (mod 4), for each $x \in \{0,1,2,\ldots,7\}$. Then φ is a ring homomorphism. For instance,

$$\varphi(a \odot b) = \varphi(a) \odot \varphi(b),$$

where the multiplication on the left is carried out mod 8 and that on the right is carried out mod 4. For if $\overline{}$ denotes remainders mod 4, this equation says

$$\overline{\text{remainder of } ab \text{ (mod 8)}} = \bar{a} \cdot \bar{b} .$$

The left-hand side equals $\overline{ab}$, since the remainder of ab (mod 8) differs from ab by a multiple of 4 (in fact, by a multiple of 8). Similarly, the right-hand side equals $\overline{ab}$, so the equation is true.

2. The mapping $\varphi:(\mathbb{Z}, +, \cdot) \to (\mathbb{Z}, +, \cdot)$ given by $\varphi(n) = 2n$ is an additive group homomorphism, but it is not a ring homomorphism since $\varphi(ab) = 2ab$ and $\varphi(a)\varphi(b) = 4ab$.

3. Let R be the ring of all real numbers of the form $a + b\sqrt{2}$, where $a, b \in \mathbb{Z}$. Map $R \to R$ by $\varphi(a + b\sqrt{2}) = a - b\sqrt{2}$. It is clear that φ preserves sums, and φ preserves products since

$$\varphi[(a + b\sqrt{2})(c + d\sqrt{2})] = \varphi[ac + 2bd + (bc + ad)\sqrt{2}]$$
$$= ac + 2bd - (bc + ad)\sqrt{2} = (a - b\sqrt{2})(c - d\sqrt{2})$$
$$= \varphi(a + b\sqrt{2})\varphi(c + d\sqrt{2}).$$

Thus φ is a ring homomorphism. It is in fact an isomorphism from R onto itself, that is, an **automorphism** of R.

4. Let $\varphi: M_2(\mathbb{R}) \to (\mathbb{R}, +, \cdot)$ be given by $\varphi\left[\begin{pmatrix} a & b \\ c & d \end{pmatrix}\right] =$ the determinant of $\begin{pmatrix} a & b \\ c & d \end{pmatrix}$. Then φ preserves products, but not sums; for instance,

$$\varphi\left[\begin{pmatrix} 1 & 0 \\ 0 & 0 \end{pmatrix} + \begin{pmatrix} 0 & 0 \\ 0 & 1 \end{pmatrix}\right] = \varphi\left[\begin{pmatrix} 1 & 0 \\ 0 & 1 \end{pmatrix}\right] = 1 \neq 0 = \varphi\left[\begin{pmatrix} 1 & 0 \\ 0 & 0 \end{pmatrix}\right] + \varphi\left[\begin{pmatrix} 0 & 0 \\ 0 & 1 \end{pmatrix}\right].$$

Thus φ is not a ring homomorphism.

5. Let $\mathbb{C}$ be the field of complex numbers and $\mathbb{H}$ the ring of quaternions. Let $\varphi: \mathbb{C} \to \mathbb{H}$ be given by

$$\varphi(a + bi) = aI + bJ.$$

φ preserves sums, and it preserves products since

$$\varphi[(a + bi)(c + di)] = \varphi(ac - bd + (ad + bc)i)$$
$$= (ac - bd)I + (ad + bc)J = (aI + bJ)(cI + dJ)$$
$$= \varphi(a + bi)\varphi(c + di).$$

φ is a monomorphism from $\mathbb{C}$ into $\mathbb{H}$, and an isomorphism from $\mathbb{C}$ onto a subring of $\mathbb{H}$.

6. Let R be the ring of real-valued functions on the real line, under pointwise addition and multiplication. R has a subring isomorphic to $(\mathbb{R}, +, \cdot)$—in fact, zillions of them! For let r be any real number; then $S = \{ f \in R \mid f(x) = 0 \text{ for every } x \neq r \}$ is a subring of R and is isomorphic to $\mathbb{R}$ via the mapping given by "evaluating at r," that is, by $\varphi(f) = f(r)$. For instance, if $f, g \in S$ we have

$$\varphi(f + g) = (f + g)(r) = f(r) + g(r) = \varphi(f) + \varphi(g),$$

and likewise for products. φ is one-to-one and onto because for each $a \in \mathbb{R}$ there is exactly one $f \in S$ such that $f(r) = a$.

Similarly, R has many subrings isomorphic to $\mathbb{R} \oplus \mathbb{R}$. For example, if r and s are distinct real numbers then $\{ f \in R \mid f(x) = 0 \text{ if } x \text{ is not } r \text{ or } s \}$ will do.

The next two theorems are the translations of Theorems 12.1 and 12.6 into the context of rings. From now on we will usually say "homomorphism" instead of "ring homomorphism" if it is clear that we mean ring homomorphism and not just group homomorphism.

THEOREM 18.3 Let R, S, T be rings and let $\varphi: R \to S$ and $\psi: S \to T$ be homomorphisms. Then:

i) $\psi \circ \varphi: R \to T$ is a homomorphism;

ii) if φ and ψ are both isomorphisms, so is $\psi \circ \varphi$;

iii) if φ is an isomorphism, so is $\varphi^{-1}: S \to R$.

THEOREM 18.4 Let $\varphi: R \to T$ be a homomorphism. Then:

i) if S is a subring of R, then $\varphi(S)$ is a subring of T;

ii) if U is a subring of T, then $\varphi^{-1}(U)$ is a subring of R;

iii) if U is an ideal of T, then $\varphi^{-1}(U)$ is an ideal of R;

iv) if φ is *onto* and S is an ideal of R, then $\varphi(S)$ is an ideal of T.

An easy example shows that the assumption that φ is onto cannot be dropped in part (iv). Let $\varphi: \mathbb{Z} \to \mathbb{R}$ be the identity map. $\mathbb{Z}$ is certainly an ideal of $\mathbb{Z}$, but $\varphi(\mathbb{Z}) = \mathbb{Z}$ is not an ideal of $\mathbb{R}$.

The relationship between homomorphisms and normal subgroups in group theory has its counterpart in a parallel relationship between homomorphisms and ideals in ring theory. If I is an ideal of R, then the mapping $\rho: R \to R/I$ given by $\rho(r) = I + r$ is a (ring) homomorphism, called the **canonical homomorphism**. Note that, in terms of just the additive groups $(R, +)$ and $(R/I, +)$, ρ is the canonical group homomorphism. ρ preserves products since

$$\rho(rs) = I + rs = (I + r)(I + s) = \rho(r)\rho(s).$$

We can recover I from ρ by forming the set $\{ r \in R \mid \rho(r) = 0_{R/I} \}$. In general, if we are given a ring homomorphism $\varphi: R \to S$, we define

$$\ker(\varphi) = \{ r \in R \mid \varphi(r) = 0_S \}.$$

Ker(φ) is just the kernel of φ considered as a homomorphism of abelian groups, and it is therefore clear that ker(φ) is an additive subgroup of R. Furthermore, if $r \in$ ker(φ) and $x \in R$, then xr and rx are both in ker(φ), because, for instance, $\varphi(xr) = \varphi(x)\varphi(r) = \varphi(x) \cdot 0_S = 0_S$. Hence ker($\varphi$) is an ideal.

The following result rounds out the picture by telling us to what extent φ can be recovered from ker(φ).

THEOREM 18.5 (Fundamental theorem on ring homomorphisms) If $\varphi: R \to T$ is an onto homomorphism, then $R/\text{ker}(\varphi) \cong T$. Moreover, if ρ is the canonical homomorphism from R onto $R/\text{ker}(\varphi)$, then there is an isomorphism $\bar{\varphi}: R/\text{ker}(\varphi) \to T$ such that $\bar{\varphi} \circ \rho = \varphi$.

PROOF. Just as for groups: Define $\bar{\varphi}: R/\text{ker}(\varphi) \to T$ by

$$\bar{\varphi}(\text{ker}(\varphi) + r) = \varphi(r).$$

This mapping is well defined, one-to-one, onto, and a homomorphism of additive groups by the proof of the Fundamental Theorem on Group Homomorphisms. It preserves products since

$$\bar{\varphi}([\text{ker}(\varphi) + r][\text{ker}(\varphi) + s]) = \bar{\varphi}(\text{ker}(\varphi) + rs) = \varphi(rs)$$
$$= \varphi(r)\varphi(s) = \bar{\varphi}(\text{ker}(\varphi) + r)\bar{\varphi}(\text{ker}(\varphi) + s). \qquad \square$$

If $\varphi: R \to T$ is an onto homomorphism, then we have a one-to-one correspondence between the subrings of T and the subrings of R that contain ker(φ), with ideals corresponding to ideals. Namely, to a subring $S \supseteq \text{ker}(\varphi)$ we associate the subring $\varphi(S)$ of T. It is easy to check that this establishes a one-to-one correspondence—everything goes just as it did for groups. If S is an ideal, then $\varphi(S)$ is an ideal because φ is onto; conversely, if $\varphi(S)$ is an ideal, then S is an ideal because $S = \varphi^{-1}(\varphi(S))$, as follows from the fact that $S \supseteq \text{ker}(\varphi)$.

Examples

1. Let $n \geqslant 1$, and let $\varphi: (\mathbb{Z}, +, \cdot) \to (\mathbb{Z}_n, \oplus, \odot)$ be given by $\varphi(x) = \bar{x}$, the remainder of x (mod n). Then φ is an onto homomorphism, and ker(φ) $= n\mathbb{Z}$, so

$$(\mathbb{Z}, +, \cdot)/n\mathbb{Z} \cong (\mathbb{Z}_n, \oplus, \odot).$$

The distinct subrings of $(\mathbb{Z}_n, \oplus, \odot)$ are $\langle d_1 \rangle, \ldots, \langle d_k \rangle$, where $d_1, \ldots, d_k$ are the positive divisors of n (Corollary 5.6). The distinct subrings of $(\mathbb{Z}, +, \cdot)$ that contain $n\mathbb{Z}$ are $d_1\mathbb{Z}, d_2\mathbb{Z}, \ldots, d_k\mathbb{Z}$, so the correspondence between the subrings is very transparent in this case.

2. Let $\varphi: \mathbb{Z} \oplus \mathbb{Z} \to \mathbb{Z}$ be given by $\varphi[(m,n)] = n$. Then φ is onto and $\ker(\varphi)$ is the ideal $\{(m,0)|m \in \mathbb{Z}\}$. We have

$$(\mathbb{Z} \oplus \mathbb{Z})/\ker(\varphi) \cong \mathbb{Z}.$$

The ideals of $\mathbb{Z} \oplus \mathbb{Z}$ that contain $\ker(\varphi)$ correspond in an obvious way to the ideals of $\mathbb{Z}$.

3. Let R be the ring of real-valued functions on $\mathbb{R}$, under pointwise operations. Fix some real number r in mind, and define a mapping $\varphi_r: R \to \mathbb{R}$ by

$$\varphi_r(f) = f(r).$$

φ_r is an onto homomorphism, and $\ker(\varphi_r) = I_r = \{f \in R | f(r) = 0\}$. Since

$$R/\ker(\varphi_r) \cong \mathbb{R}$$

and $\mathbb{R}$ is a field, Theorem 17.7 tells us that I_r is maximal in R for every r. We verified this by more heavy-handed means in Example 3 on p. 168.

The proofs of the second and third isomorphism theorems offer no surprises, so we leave them as easy exercises.

THEOREM 18.6 (Second isomorphism theorem for rings) Let S be a subring of R and let I be an ideal. Then $S \cap I$ is an ideal of S, and

$$S/(S \cap I) \cong (S+I)/I.$$

Here $S + I$ is the subring $\{s + x | s \in S, \ x \in I\}$ of R.

THEOREM 18.7 (Third isomorphism theorem for rings) Let I and J be ideals of R and suppose $I \subseteq J$. Then J/I is an ideal of R/I and

$$\frac{R/I}{J/I} \cong R/J.$$

With these theorems stated, we have now remodeled all our basic machinery so that it is suitable for use in ring theory. We will proceed to some new results, each of which involves ring homomorphisms.

We first want to show that every field F has a subfield that is isomorphic either to $\mathbb{Q}$ or to $(\mathbb{Z}_p, \oplus, \odot)$ for some prime p. In fact, more is true: F has exactly one such subfield. We are interested in this result because it provides a useful classification of fields. The nature of the solutions of equations involving elements of F can be heavily influenced by whether it is $\mathbb{Q}$, on the one hand, or some $\mathbb{Z}_p$, on the other, which is isomorphic to a subfield of F.

THEOREM 18.8 Let F be a field. Then F has a subfield that is isomorphic to either $\mathbb{Q}$ or some $\mathbb{Z}_p$, and F has only one such subfield.

PROOF. We split the proof into "existence" and "uniqueness" pieces.

Existence. We shall be working with integers, so let us denote the multiplicative identity of F by e, rather than 1, to avoid confusion. The subfield we seek will consist of e and just those additional elements that are required to form a field. We could obtain it by brute force, but we prefer to use a little finesse.

Define a homomorphism $\varphi: \mathbb{Z} \to F$ by $\varphi(n) = ne$, and let S be the image of φ. Since S contains e and is a subring of F, S is a domain. Therefore, since $\mathbb{Z}/\ker(\varphi) \cong S$, Corollary 17.6 tells us that $\ker(\varphi)$ is a proper prime ideal of $\mathbb{Z}$. Hence $\ker(\varphi)$ is either $p\mathbb{Z}$, for some prime p, or the trivial ideal $\{0\}$. If $\ker(\varphi) = p\mathbb{Z}$, then $S \cong \mathbb{Z}/p\mathbb{Z} \cong \mathbb{Z}_p$, so S is the subfield we seek. If $\ker(\varphi) = \{0\}$, then φ is an isomorphism from $\mathbb{Z}$ onto S, and we get an isomorphism from $\mathbb{Q}$ onto a subfield of F by defining $\varphi(m/n) = me/ne$ [that is, $(me)(ne)^{-1}$]. This is a well-defined one-to-one mapping, since for integers m, n, m', n' with $n, n' \neq 0$, we have $m/n = m'/n'$ iff $mn' = nm'$ iff $mn'e = nm'e$ iff $(me)(n'e) = (ne)(m'e)$ iff $me/ne = m'e/n'e$. It is easy to check that φ is a homomorphism.

Uniqueness. Note first that if K is any subfield of F, then the multiplicative identity of K must be the same as e. This is so because $K - \{0_F\}$ is a subgroup of the multiplicative group $F - \{0_F\}$, hence has the same identity element.

It follows that if there is an isomorphism $\psi: K \to \mathbb{Z}_p$, then $pe = 0_F$, while if there is an isomorphism $\psi: K \to \mathbb{Q}$, then e has infinite order in $(F, +)$. Therefore, if F has a subfield isomorphic to some $\mathbb{Z}_p$, F can have neither a subfield isomorphic to some $\mathbb{Z}_q$, $q \neq p$, nor a subfield isomorphic to $\mathbb{Q}$.

To finish the proof, we must show that F cannot have two subfields isomorphic to $\mathbb{Q}$ or to the same $\mathbb{Z}_p$. But if $\psi: K \to \mathbb{Q}$ is an isomorphism, then $\psi(e) = 1$, and since $\mathbb{Q} = \{m/n \mid m, n \in \mathbb{Z}, n \neq 0\}$, it follows that $K = \{me/ne \mid m, n \in \mathbb{Z}, n \neq 0\}$. Likewise, any subfield isomorphic to $\mathbb{Z}_p$ must be the set $\{e, 2e, \ldots, (p-1)e\}$. $\square$

The subfield of F that is isomorphic to $\mathbb{Q}$ or some $\mathbb{Z}_p$ is called the **prime subfield** of F. If F is itself either $\mathbb{Q}$ or some $\mathbb{Z}_p$, then F is its own prime subfield, and $\mathbb{Q}$ and the $\mathbb{Z}_p$'s are accordingly called **prime fields**.

If the prime subfield of F is isomorphic to $\mathbb{Q}$, equivalently if 1_F has infinite order in $(F, +)$, we say that F is **of characteristic 0**. Thus $\mathbb{Q}$, $\mathbb{R}$, and $\mathbb{C}$ are all of characteristic 0. If the prime subfield of F is isomorphic to $\mathbb{Z}_p$, we say that F is **of characteristic p**. It is clear, for example, that every finite field must be of characteristic p for some prime p.

Our next two results are examples of what are called **embedding theorems**, in that they state that a ring of some kind can be embedded in a ring of some other kind. To say that R can be embedded into S is simply to say that there is an embedding (monomorphism) from R into S, that is, R is isomorphic to a subring of S. The usual purpose of an embedding theorem is to remedy some defect of a given ring by embedding the ring in a bigger ring, where things are better. For example, we know that some rings lack multiplicative identities. But:

THEOREM 18.9 Let R be a ring. Then R can be embedded in a ring with unity.

PROOF. Let $S = \{(r,n)|r \in R,\ n \in \mathbb{Z}\}$, with operations defined by

$$(r,n) + (s,m) = (r+s, n+m)$$

and

$$(r,n)(s,m) = (rs + ns + mr, nm).$$

It is clear that S forms an abelian group under addition, and a routine calculation shows that multiplication is associative and the distributive laws hold. Thus S is a ring.

We map $R \to S$ by $\varphi(r) = (r,0)$. φ is clearly one-to-one, and φ is a homomorphism because

$$\varphi(r+s) = (r+s, 0) = (r,0) + (s,0) = \varphi(r) + \varphi(s),$$

and

$$\varphi(rs) = (rs, 0) = (r,0)(s,0) = \varphi(r)\varphi(s).$$

S has unity $(0_R, 1)$, since

$$(r,n)(0_R, 1) = (r \cdot 0_R + n0_R + 1r, n) = (r,n)$$

and

$$(0_R, 1)(r,n) = (0_R \cdot r + 1r + n0_R, n) = (r,n). \quad \square$$

Notice that if R already had a unity 1_R to begin with, then 1_R is mapped to $(1_R, 0)$, which is not the unity in S. In this case the subring $\{(r,0)|r \in R\}$ of S has a unity different from that of S.

Our next result is motivated by a desire to provide a multiplicative inverse for every nonzero element of a commutative ring R with unity, that is, to embed R into a field. Of course this is not always possible, for if R is going to be isomorphic to a subring of a field, then R cannot contain any nonzero zero-divisors. Thus the best we could hope for would be that every *integral domain* could be embedded in a field. And this is precisely what we get.

THEOREM 18.10 If D is an integral domain, then D can be embedded in a field.

PROOF. The field we seek must at least have elements corresponding to a/b (that is, ab^{-1}), for all $a,b \in D$, $b \neq 0_D$. We attempt to build it by starting with the set $S = \{(a,b)|a,b \in D,\ b \neq 0_D\}$, in the hope that (a,b) will turn out to be the element corresponding to a/b.

One problem arises immediately, in that if (a,b) is going to correspond to a/b, then for $x \neq 0_D$, (xa, xb) should correspond to xa/xb, which should be the same element as a/b. In other words, S contains more than one contender for the role of a/b. We overcome this difficulty by introducing an equivalence relation on S, so as to lump all these contenders together into one object.

Define $(a,b)R(c,d)$ iff $ad = bc$ (or, intuitively, $a/b = c/d$). R is reflexive because $(a,b)R(a,b)$ says $ab = ba$, which is true because D is commutative. If $(a,b)R(c,d)$, then $ad = bc$, and so $cb = da$, and $(c,d)R(a,b)$. Thus R is symmetric. Finally, if $(a,b)R(c,d)$ and $(c,d)R(e,f)$, then

$$ad = bc \qquad \text{and} \qquad cf = de; \qquad \text{[18.1]}$$

we wish to conclude that $af = be$, that is, $(a,b)R(e,f)$. Multiplying the equations in [18.1] yields $adcf = bcde$, whence $acf = bce$ since $d \neq 0_D$, and D is a domain. If $c \neq 0_D$ we get $af = be$, as desired. If $c = 0_D$ then [18.1] gives us $a = 0_D$ and $e = 0_D$, so again $af = be$.

Now let F be the set of equivalence classes under R:

$$F = \{ \overline{(a,b)} \mid (a,b) \in S \}.$$

We introduce addition and multiplication on F by defining

$$\overline{(a,b)} + \overline{(c,d)} = \overline{(ad + bc, bd)}$$

[because we want $a/b + c/d = (ad + bc)/bd$], and

$$\overline{(a,b)} \cdot \overline{(c,d)} = \overline{(ac, bd)}$$

[because we want $a/b \cdot c/d = ac/bd$]. In both cases we use the fact that $bd \neq 0_D$ to guarantee that the answer is in F. Both operations are well defined in that the choice of representatives for the classes involved does not affect the answers (Exercise 18.28).

We claim now that F is a field. First we show that F is an abelian group under addition. Associativity demands that

$$\left[\overline{(a,b)} + \overline{(c,d)} \right] + \overline{(e,f)} = \overline{(a,b)} + \left[\overline{(c,d)} + \overline{(e,f)} \right].$$

This reduces to

$$\overline{(ad + bc, bd)} + \overline{(e,f)} = \overline{(a,b)} + \overline{(cf + de, df)}, \qquad \text{or}$$
$$\overline{((ad + bc)f + (bd)e, (bd)f)} = \overline{(a(df) + b(cf + de), b(df))},$$

which we know is true because we have the same element of S on both sides. The additive identity is $\overline{(0,b)}$, for any nonzero $b \in D$:

$$\overline{(c,d)} + \overline{(0,b)} = \overline{(cb + d0, db)} = \overline{(c,d)}.$$

[Note that the choice of b does not affect the class of $(0,b)$.] The inverse of

$\overline{(c,d)}$ is $\overline{(-c,d)}$, and addition is commutative since

$$\overline{(a,b)} + \overline{(c,d)} = \overline{(ad+bc,bd)}$$
$$= \overline{(cb+da,db)} = \overline{(c,d)} + \overline{(a,b)}.$$

To complete the verification that F is a field, we must show that multiplication is associative and commutative (but these are obvious), that the distributive laws hold (Exercise 18.29), that F has a multiplicative identity distinct from the additive identity [$\overline{(b,b)}$ is it, for any nonzero $b \in D$], and that every nonzero element of F is a unit. For this last step, note that if $\overline{(c,d)} \neq 0_F$, then $c \neq 0_D$, so $\overline{(d,c)} \in F$ and

$$\overline{(c,d)} \cdot \overline{(d,c)} = \overline{(cd,dc)} = 1_F.$$

We have now concocted the field F, and all that remains is to embed D in F. The way is clear: let

$$\varphi(d) = \overline{(d,1_D)}.$$

φ is one-to-one since if $c \neq d$ then $\overline{(c,1)} \neq \overline{(d,1)}$. φ is a homomorphism since

$$\varphi(c+d) = \overline{(c+d,1)} = \overline{(c,1)} + \overline{(d,1)} = \varphi(c) + \varphi(d),$$

and

$$\varphi(cd) = \overline{(cd,1)} = \overline{(c,1)} \cdot \overline{(d,1)} = \varphi(c)\varphi(d).$$

This completes the proof. $\square$

The field constructed in this proof is called the **field of quotients of D**, or, more briefly, the **quotient field of D**. Note that every element of the quotient field can be written in the form

$$\overline{(a,b)} = \overline{(a,1_D)}\ \overline{(1_D,b)} = \varphi(a)[\varphi(b)]^{-1},$$

for some $a,b \in D$, $b \neq 0_D$.

This embedding theorem (or any embedding theorem, for that matter) becomes more convenient to use if we modify it in the following way. Let $\varphi: R \to S$ be an embedding. We construct a ring S' that is isomorphic to S and actually contains R (rather than just some ring isomorphic to R) as a subring. Let S' be obtained from S by replacing $\varphi(r)$ by r, for each $r \in R$:

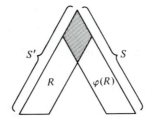

The shaded portion is $S - \varphi(R) = S' - R$.

Let ψ be the following one-to-one onto mapping from S' to S:

$$\psi(x) = \begin{cases} \varphi(x) & \text{if } x \in R, \\ x & \text{if } x \in S' - R. \end{cases}$$

We define addition and multiplication on S' by

$$x + y = \psi^{-1}(\psi(x) + \psi(y)), \quad \text{and}$$
$$xy = \psi^{-1}(\psi(x)\psi(y))$$

[18.2]

for all $x, y \in S'$. (The operations on the right are carried out in the known ring S. Note that if x and y are both in R, then the operations on the left coincide with the original operations in R.) The equations in [18.2] give us

$$\psi(x + y) = \psi(x) + \psi(y)$$

and

$$\psi(xy) = \psi(x)\psi(y).$$

Thus ψ is an isomorphism from S' onto S, and therefore S' is a ring and is isomorphic to S.

Since R is a subring of S', we say that S' is an **extension** of R. Thus Theorem 18.9 tells us that any ring actually has an extension which is a ring with unity, and Theorem 18.10 says that any integral domain has an extension which is a field. In particular, we can regard the quotient field of a domain D as an extension of D, and we thus obtain the following sharper version of Theorem 18.10.

THEOREM 18.11 Let D be a domain. Then D has an extension F which is a field, such that every element of F can be written in the form a/b for some $a, b \in D$, $b \neq 0_D$.

It can be shown that if F_1 and F_2 are fields, both having the properties described in Theorem 18.11, then there is an isomorphism $\varphi: F_1 \rightarrow F_2$ such that $\varphi(d) = d$ for every $d \in D$ (see Exercise 18.30). For this reason, any such field is called a quotient field of D.

In particular, the quotient field of $\mathbb{Z}$ is $\mathbb{Q}$. Thus one very concrete benefit of the proof of Theorem 18.10 is that it shows us how to construct the rational numbers from the integers.

EXERCISES

18.1 Which of the following are ring homomorphisms?

a) $\varphi: \mathbb{R} \rightarrow \mathbb{R}$ by $\varphi(x) = |x|$

b) $\varphi: \mathbb{C} \rightarrow \mathbb{C}$ by $\varphi(a + bi) = a - bi$

c) $\varphi: \mathbb{C} \rightarrow \mathbb{R}$ by $\varphi(a + bi) = a$

d) Let $R = \{a + b\sqrt{2} \mid a, b \in \mathbb{Z}\}$ and $S = \{a + b\sqrt{3} \mid a, b \in \mathbb{Z}\}$. Let $\varphi: R \to S$ be given by $\varphi(a + b\sqrt{2}) = a + b\sqrt{3}$.

e) Let R be the ring of polynomials with real coefficients, and let $\varphi: R \to R$ be given by $\varphi(p(x)) = p'(x)$, the derivative of $p(x)$.

18.2 Consider the mapping $\varphi: \mathbb{Z}_3 \to \mathbb{Z}_6$ given by $\varphi(x) = 2x$, for $x = 0, 1, 2$. Is φ a ring homomorphism? How about the mapping $\varphi(x) = $ remainder of $4x \pmod 6$?

18.3 Determine whether the rings $2\mathbb{Z}$ and $3\mathbb{Z}$ are isomorphic.

18.4 Let X be a nonempty set. Let $R = (P(X), \triangle, \cap)$. For each subset Y of X, define a function $f_Y: X \to (\mathbb{Z}_2, \oplus, \odot)$ by

$$f_Y(x) = \begin{cases} 1 & \text{if } x \in Y, \\ 0 & \text{if } x \notin Y. \end{cases}$$

Show that the set $\{f_Y \mid Y \subseteq X\}$ forms a ring of functions on X, under pointwise addition and multiplication (mod 2), and that this ring is isomorphic to R.

18.5 Find all homomorphisms from $(\mathbb{Z}, +, \cdot)$ into itself.

18.6 Show that the ring $\{a + b\sqrt{2} \mid a, b \in \mathbb{Z}\}$ has precisely two automorphisms.

18.7 What are all the quotient rings of $\mathbb{Z}_n$?

18.8 Prove Theorem 18.1 (iii) and (iv).

18.9 Prove Theorem 18.2 (iii).

18.10 Prove Theorem 18.3.

18.11 Prove Theorem 18.4.

18.12 Suppose F is a field. Determine all the homomorphic images of F, up to isomorphism. (That is, determine for which rings R there exists an onto homomorphism $\varphi: F \to R$.)

18.13 a) Let $\varphi: R \to S$ be a homomorphism, and let I be an ideal of R such that $I \subseteq \ker(\varphi)$. Show that φ induces a homomorphism from R/I into S.

b) Let $\varphi: R \to S$ be a homomorphism, let I be an ideal of R, and let J be an ideal of S such that $\varphi(I) \subseteq J$. Show that φ induces a homomorphism from R/I into S/J.

c) Let $\varphi: R \to S$ be an isomorphism, and let I be an ideal of R. Show that φ induces an isomorphism from R/I onto $S/\varphi(I)$.

18.14 Let $\varphi: R \to S$ be a homomorphism, and suppose S has a unity, 1_S. Show that $\varphi^{-1}(\{1_S\})$ is an ideal in R iff S is trivial.

18.15 Let $\varphi: R \to S$ be a homomorphism. Show that φ is one-to-one iff $\ker(\varphi) = \{0_R\}$.

18.16 Let R be a ring, and let I and J be ideals of R such that $I + J = R$. (See Exercise 17.33 for the definition of $I + J$.) Show that $R/(I \cap J) \cong R/I \oplus R/J$.

18.17 Let I be an ideal of R and J an ideal of S. Show that

$$(R \oplus S)/(I \oplus J) \cong R/I \oplus S/J.$$

18.18 Let I and J be ideals of R such that $I + J = R$ and $I \cap J = \{0\}$. Show that $R/I \cong J$.

18.19 Suppose R has ideals I and J such that $I + J = R$ and $I \cap J = \{0\}$. Show that $R \cong I \oplus J$. Generalize to more than two summands.

18.20 Prove Theorem 18.6.

18.21 Prove Theorem 18.7.

18.22 Let $\varphi: R \to S$ be a homomorphism.
   a) Show that if J is a prime ideal in S then $\varphi^{-1}(J)$ is a prime ideal in R.
   b) Show that if I is a prime ideal in R, then $\varphi(I)$ need not be a prime ideal in S, even if φ is onto. Show that $\varphi(I)$ will be prime, however, if φ is onto and $\ker(\varphi) \subseteq I$.

18.23 Let $\varphi: R \to S$ be an onto homomorphism.
   a) Show that if J is a maximal ideal in S then $\varphi^{-1}(J)$ is a maximal ideal in R.
   b) Show that if I is a maximal ideal in R, then $\varphi(I)$ is maximal iff it is proper, and that it is maximal if $\ker(\varphi) \subseteq I$.

18.24 a) Let F be a field and let $\text{Aut}(F)$ denote the set of ring automorphisms of F. Show that $\text{Aut}(F)$ forms a group under composition of functions.
   b) Let K be a subfield of F. Show that the set
$$\{\varphi \in \text{Aut}(F) \,|\, \varphi(k) = k \text{ for every } k \in K\}$$
is a subgroup of $\text{Aut}(F)$.

18.25 Let R be a ring. Does the set of automorphisms of R form a ring under addition and composition?

18.26 Prove that multiplication is associative and the distributive laws hold in the ring S of Theorem 18.9.

18.27 Prove that if R is a commutative ring, then R can be embedded in a commutative ring with unity.

18.28 Prove that the operations introduced in the proof of Theorem 18.10 are well defined.

18.29 Prove the distributive laws for the field F of Theorem 18.10.

18.30 a) Let D be a domain, and let F be a field extending D, such that every element of F can be written in the form ab^{-1}, with $a, b \in D$. Let K be a field, and let $\varphi: D \to K$ be an embedding. Show that there is a unique embedding $\psi: F \to K$ such that $\psi(d) = \varphi(d)$ for every $d \in D$. (We say that ψ *extends* φ. In the sense of this result, the quotient field of D is the smallest field extending D.)
   b) In particular, let K be a field extending D, and suppose K also has the property that each of its elements can be written as ab^{-1}, with $a, b \in D$. Show that there is an isomorphism ψ from F onto K such that $\psi(d) = d$ for every $d \in D$.

18.31 A field F is said to be **orderable** if F has a subset F^+ with the following properties:

i) F^+ is closed under addition and multiplication;

ii) For every $x \in F$ exactly one of the following holds:
$$-x \in F^+, \qquad x = 0_F, \qquad \text{or} \quad x \in F^+.$$

If F is orderable, and we have chosen a subset F^+ with the indicated properties, then F is said to be **ordered**, and F^+ is called the set of **positive** elements of F.

a) Suppose that F is ordered, with set of positive elements F^+. Show that $x^2 \in F^+$ for every nonzero $x \in F$. Show that, in particular, $1_F \in F^+$.

b) Show that every orderable field has characteristic 0.

c) Show that $\mathbf{C}$ is not orderable.

18.32 Let F be an ordered field with set of positive elements F^+. For $x, y \in F$, define $x < y$ (equivalently, $y > x$) to mean that $y - x \in F^+$. Show that, for all $x, y, z \in F$:

a) exactly one of the following holds:
$$x < y, \qquad x = y, \qquad \text{or} \quad x > y;$$

b) if $x < y$ and $y < z$, then $x < z$;

c) if $x > 0$ and $y > 0$, then $x + y > 0$ and $xy > 0$;

d) if $x < y$, then $x + z < y + z$;

e) if $x < y$ and $z > 0$, then $xz < yz$, while if $x < y$ and $z < 0$, then $xz > yz$.

POLYNOMIALS

We have all been familiar with polynomials since our days in high school algebra. We are by now accustomed to thinking of them as functions of the form $f(x) = a_0 + a_1 x + \cdots + a_n x^n$, with x being a variable and the a_i's being real constants. In this section we will take another look at them, from the point of view of abstract algebra.

We shall denote variables by upper-case letters: $X, Y, \ldots$. If R is a ring, then by a **polynomial in X with coefficients from R** we mean an infinite formal symbol

$$a_0 + a_1 X + a_2 X^2 + a_3 X^3 + \cdots,$$

where each $a_i \in R$ and there is some n such that $a_i = 0$ for all $i > n$. The a_i's are called the **coefficients** of the polynomial, and a_i is called the coefficient of X^i. If $a_n \neq 0$ and $a_i = 0$ for all $i > n$, then we usually write the above polynomial more simply as

$$a_0 + a_1 X + a_2 X^2 + \cdots + a_n X^n,$$

but we still regard it as having a coefficient a_j for every $j \geqslant 0$. This approach is often very convenient. For instance, it makes it easy for us to say what we mean by two polynomials being equal. If

$$f(X) = a_0 + a_1 X + a_2 X^2 + \cdots \qquad \text{and} \qquad g(X) = b_0 + b_1 X + b_2 X^2 + \cdots$$

are polynomials with coefficients from R, then we say that they are *equal*, and we write $f(X) = g(X)$, if $a_i = b_i$ for every i. Notice that we don't have to worry about one of $f(X)$ or $g(X)$ having more coefficients than the other (for instance, extra 0-coefficients) because we regard them both as having a coefficient for each power of X.

Observe that a polynomial $f(X)$ with coefficients from R gives us a function on R, obtained by plugging elements of R in for X and interpreting addition and multiplication as the operations in R. However, our point of view at the moment is that $f(X)$ is a formal expression, and *not* the function this expression induces on R. This distinction is a significant one, because it is possible for two different polynomials to induce the same function on R. For example, if R is $(\mathbb{Z}_2, \oplus, \odot)$, then both $0 + 0X + 0X^2 + \cdots$ and $0 + 1X + 1X^2$ give us the function that maps every element of R to 0.

In denoting polynomials, we usually omit terms with coefficient 0, wherever they occur. For instance, $a_0 + 0X + a_2X^2$ can be written as $a_0 + a_2X^2$. On the other hand, if R is a ring with unity, then we usually do not bother to write 1_R when it occurs as a coefficient. Thus if we are considering polynomials with coefficients from $\mathbb{Z}_2$, then we write $X + X^2$ in place of $1X + 1X^2$.

We denote the set of all polynomials in X with coefficients from R by $R[X]$. The elements of $R[X]$ are also sometimes called polynomials *over* R.

We turn $R[X]$ into a ring by introducing the natural addition and multiplication. If

$$f(X) = a_0 + a_1X + a_2X^2 + \cdots \quad \text{and} \quad g(X) = b_0 + b_1X + b_2X^2 + \cdots,$$

then we add $f(X)$ and $g(X)$ by adding corresponding coefficients:

$$f(X) + g(X) = (a_0 + b_0) + (a_1 + b_1)X + (a_2 + b_2)X^2 + \cdots.$$

We multiply just as we did in high school algebra: $f(X)g(X) = c_0 + c_1X + c_2X^2 + \cdots$, where for each n,

$$c_n = a_0b_n + a_1b_{n-1} + a_2b_{n-2} + \cdots + a_{n-1}b_1 + a_nb_0.$$

Thus the product is obtained by multiplying everything out, using the rule $aX^i \cdot bX^j = abX^{i+j}$, and then collecting terms that involve the same power of X. The product is indeed a polynomial over R, because there exists n such that $c_i = 0$ for every $i > n$. Specifically, if $a_i = 0$ for all $i > m$ and $b_i = 0$ for all $i > l$, then we can take $n = m + l$. For with this choice of n, each term a_ib_j in the expression for c_k, $k > n$, must have $i + j > m + l$, so either $i > m$ or $j > l$, and therefore either $a_i = 0$ or $b_j = 0$, whence $a_ib_j = 0$.

Observe that, under these definitions, a polynomial such as $a_0 + a_1X + a_2X^2$ is actually the sum of a_0, a_1X, and a_2X^2 in $R[X]$. By the same token, if R is a ring with unity, so that $X \in R[X]$, then a_2X^2 is the product of a_2X and X.

It is not very difficult (but neither is it very exciting) to verify that $R[X]$ does form a ring under the given operations. We shall forego getting into this, and instead take up the more interesting question of which properties of R carry over to the new ring $R[X]$.

Some things can be seen at once. For instance, if R has unity 1, then $R[X]$ has unity $1+0X+0X^2+\cdots$, which we also denote by 1. If R is commutative, then so is $R[X]$, for if

$$f(X)=a_0+a_1X+a_2X^2+\cdots \quad \text{and} \quad g(X)=b_0+b_1X+b_2X^2+\cdots,$$

then the coefficient of X^n in $f(X)g(X)$ is

$$a_0b_n+a_1b_{n-1}+\cdots+a_{n-1}b_1+a_nb_0,$$

and the coefficient of X^n in $g(X)f(X)$ is

$$b_0a_n+b_1a_{n-1}+\cdots+b_{n-1}a_1+b_na_0.$$

These are the same, by the commutativity of multiplication in R.

What if R is a domain? Must $R[X]$ be one too? The answer is yes, because if $f(X)$ and $g(X)$ are as in the last paragraph, and neither one is the zero polynomial

$$0+0X+0X^2+\cdots,$$

then we can let n be such that $a_n \neq 0$ and $a_i=0$ for all $i>n$, and let m be such that $b_m \neq 0$ and $b_i=0$ for all $i>m$. It then follows that the coefficient of X^{n+m} in $f(X)g(X)$ is a_nb_m, which is not 0 since a_n, b_m are nonzero elements of a domain. Thus $f(X)g(X)$ is not the zero polynomial, and it follows that $R[X]$ is a domain.

Arguments such as the one we have just given can be expressed more succinctly if we allow ourselves some additional terminology. If $f(X) \in R[X]$ is not the zero polynomial, then we can write

$$f(X)=a_0+a_1X+\cdots+a_nX^n, \quad \text{where } a_n \neq 0.$$

The integer n is called the **degree** of $f(X)$, and a_n is called the **leading coefficient**. We denote the degree of $f(X)$ by $\deg(f)$, or $\deg[f(X)]$. Notice that $\deg(f)=0$ iff $f(X)$ is a nonzero **constant polynomial**, that is, iff $f(X)$ is the polynomial a_0, for some nonzero $a_0 \in R$. The zero polynomial is not assigned a degree.[†]

The argument we gave above can now be expressed more precisely by saying that if R is a domain, and $f(X), g(X) \in R[X]$ have leading coefficients a_n and b_m, respectively, then $f(X)g(X)$ has leading coefficient a_nb_m. As a consequence, we have

---

[†]No confusion should result from using the same symbol to denote both $a_0 \in R$ and the constant polynomial a_0. In fact, R is isomorphic to the subring of $R[X]$ consisting of all the constant polynomials, and we often think of R itself as a subring of $R[X]$.

THEOREM 19.1 (Degree rule) If R is a domain and $f(X), g(X)$ are nonzero elements of $R[X]$, then

$$\deg\left[\,f(X)g(X)\,\right] = \deg\left[\,f(X)\,\right] + \deg\left[\,g(X)\,\right].$$

The assumption that R is a domain is crucial here. For instance, in the nondomain $\mathbb{Z}_6$, we have $(2X)(3X+1) = 2X$, or, even worse, $(2X)(3X+3) = 0$.

We have seen above that a number of properties will always be passed on from R to $R[X]$. One property that will obviously not be passed on is that of being a field. For if F is a field, then the element X is clearly not a unit in $F[X]$.

Nevertheless, it does seem reasonable to expect that assuming F to be a field will have a beneficial impact on $F[X]$, beyond that of making it a domain. Much of what can be said about $F[X]$ depends on the following analogue of the division algorithm for $\mathbb{Z}$.

THEOREM 19.2 (Division algorithm for $F[X]$) Let F be a field, and let $f(X), g(X) \in F[X]$. If $g(X) \neq 0$, then there exist $q(X), r(X) \in F[X]$ such that

$$f(X) = q(X)g(X) + r(X)$$

and either $r(X) = 0$ or $\deg(r) < \deg(g)$.

PROOF. If $f(X) = 0$, or if $f(X) \neq 0$ and $\deg(f) < \deg(g)$, we write

$$f(X) = 0 \cdot g(X) + f(X),$$

and we are done.

We now proceed by induction on $\deg(f)$. If $\deg(f) = 0$, then by the above we are done unless $\deg(g) = 0$. But in this case both $f(X)$ and $g(X)$ are constant polynomials—say, $f(X) = a_0$, $g(X) = b_0$—and thus we can write

$$f(X) = \left(a_0 b_0^{-1}\right)g(X) + 0.$$

Note that b_0^{-1} exists because F is a field and $b_0 \neq 0$.

Now assume the result has been proved for $\deg(f) < n$, and suppose

$$f(X) = a_0 + a_1 X + \cdots + a_n X^n, \quad g(X) = b_0 + b_1 X + \cdots + b_m X^m, \quad \text{with } a_n, b_m \neq 0.$$

If $n < m$ then we are done by the special case handled at the outset. Otherwise we write

$$f(X) = a_n b_m^{-1} X^{n-m} g(X) + h(X),$$

where $h(X) = 0$ or $\deg(h) < \deg(f)$. If $h(X) = 0$, we have what we want; if $h(X) \neq 0$, then the inductive hypothesis tells us that we can write

$$h(X) = q(X)g(X) + r(X), \quad \text{with } r(X) = 0 \text{ or } \deg(r) < \deg(g).$$

Thus

$$f(X) = \left[a_n b_m^{-1} X^{n-m} + q(X) \right] g(X) + r(X),$$

and we are done. $\square$

If you think about it, this proof really boils down to the "long division" of polynomials, a process you have been familiar with for years. It is not difficult to show that the quotient $q(X)$ and remainder $r(X)$ are uniquely determined by $f(X)$ and $g(X)$.

We will see in Section 20 that Theorem 19.2 has a powerful impact on the ideals of $F[X]$. At the moment, we want to use a very special case of the theorem to clarify the distinction between "polynomials as formal symbols" and "polynomials as functions." We remarked above that if F is the finite field $\mathbb{Z}_2$, then the nonzero polynomial $X^2 + X$ ($= X + X^2$) induces the zero function on F, that is, $a^2 + a = 0$ for every $a \in F$. Theorem 19.2 makes it easy to see that this kind of thing cannot happen if F is an infinite field.

If F is a field and $f(X) = a_0 + a_1 X + \cdots + a_n X^n \in F[X]$, then an element $a \in F$ is called a **root** (or **zero**) of $f(X)$ if $f(a) = 0$, that is, $a_0 + a_1 a + \cdots + a_n a^n = 0$ in F. Thus, above, each element of $\mathbb{Z}_2$ is a root of $X^2 + X$.

THEOREM 19.3 Let F be a field, $a \in F$, $f(X) \in F[X]$. Then $f(a) = 0$ iff $X - a$ divides $f(X)$ in $F[X]$ (i.e., there is $h(X) \in F[X]$ such that $f(X) = (X - a)h(X)$).

PROOF. If $f(X) = (X - a)h(X)$, then $f(a) = 0 \cdot h(a) = 0$ (see Exercise 19.12). Conversely, assume that $f(a) = 0$. By Theorem 19.2, we can write

$$f(X) = (X - a)q(X) + r(X),$$

where either $r(X) = 0$ or $\deg(r) < \deg(X - a)$. In any case, $r(X)$ must be a constant here, because $\deg(X - a) = 1$. Now since $f(a) = 0$, we have (again by Exercise 19.12)

$$0 = f(a) = (a - a)q(a) + r(a) = r(a).$$

Since $r(X)$ is constant, this means $r(X) = 0$, so $f(X) = (X - a)q(X)$. $\square$

COROLLARY 19.4 Let F be a field, $f(X)$ a polynomial of degree n over F. Then $f(X)$ has at most n distinct roots in F.

PROOF. By induction on n. If $n = 0$, then $f(X)$ is a nonzero constant polynomial, hence has no roots in F.

Now suppose $n = m + 1$, and the result is known for $n = m$. Suppose $f(X)$ has $m + 2$ distinct roots in F, say $a_1, a_2, \ldots, a_{m+2}$. If we write

$$f(X) = (X - a_1)h(X),$$

then, since F is a domain, each of $a_1, \ldots, a_{m+2}$ must be a root of either $X - a_1$ or $h(X)$. In particular, $a_2, \ldots, a_{m+2}$ are $m + 1$ distinct roots of $h(X)$. Since $h(X)$

has degree m, this contradicts the inductive hypothesis, and completes the proof. $\square$

COROLLARY 19.5 Let F be an infinite field, S an infinite subset of F. If $f(X) \in F[X]$ and $f(s) = 0$ for every $s \in S$, then $f(X)$ is the zero polynomial.

PROOF. If not, then $f(X)$ has degree n for some $n > 0$, so $f(X)$ has at most n roots and cannot have all the elements of S as roots. $\square$

COROLLARY 19.6 Let F be an infinite field, S an infinite subset of F. Suppose $f(X), g(X) \in F[X]$ and $f(s) = g(s)$ for every $s \in S$. Then $f(X) = g(X)$, that is, $f(X)$ and $g(X)$ are the same polynomial.

PROOF. Apply Corollary 19.5 to the polynomial $f(X) - g(X)$. $\square$

·Examples

1. Two polynomials in $\mathbb{R}[X]$ induce the same function on $\mathbb{R}$ iff they are the same polynomial.

2. Corollary 19.4 may fail if F is not a field. For instance, let $R = \mathbb{Z}_2 \oplus \mathbb{Z}_2$, and let $f(X) = X^2 + X \in R[X]$. Then $\deg(f) = 2$, but each of the four elements of R is a root of $f(X)$.

Corollary 19.4 (and, with it, 19.5 and 19.6) does hold true for any domain R, however. See Exercise 19.16.

3. Corollary 19.4 says that if $\deg(f) = n$, then $f(X)$ has *at most* n roots in F. It may have fewer; indeed, it may have none. For example, the polynomial $X^2 + 1$ in $\mathbb{R}[X]$ has no roots in $\mathbb{R}$.

The problem of trying to determine the roots of a polynomial in $\mathbb{R}[X]$ is one that we all encountered in our earliest experience with polynomials. One of the standard methods for dealing with this problem is, of course, to write the given polynomial as a product of simpler factors, and then to find the roots of all the factors. You may remember that in $\mathbb{R}[X]$, every nonconstant polynomial can be written as a product of factors that either have degree 1, or have degree 2 and are irreducible in the sense that they cannot be factored any further (Exercise 19.5).

DEFINITION Let F be a field. A nonconstant polynomial $f(X) \in F[X]$ is called **irreducible in $F[X]$** (or **irreducible over F**) if f cannot be written as the product of two nonconstant polynomials in $F[X]$.

In general, $F[X]$ may contain irreducible polynomials of degree higher than 2. It is still true, however, that every nonconstant polynomial can be written as a product of irreducible factors.

THEOREM 19.7 Let F be a field, and let $f(X)$ be a nonconstant polynomial in $F[X]$. Then there exist irreducible polynomials $f_1(X), \ldots, f_k(X)$ in $F[X]$ such that $f(X) = f_1(X)f_2(X) \cdots f_k(X)$.

PROOF. By induction on $\deg(f)$. If $\deg(f) = 1$, then $f(X)$ is itself irreducible, and we are done. Now suppose $\deg(f) = n$, and the theorem is proved for all polynomials of degree less than n. If $f(X)$ is irreducible, we are done. Otherwise, we can write $f(X) = g(X)h(X)$, where $g(X)$ and $h(X)$ each have degree at least 1. By the degree rule, $g(X)$ and $h(X)$ each have degree less than n, so by the inductive hypothesis we can factor $g(X)$ and $h(X)$ into irreducible factors. This yields the desired factorization of $f(X)$. $\square$

This proof of the existence of a factorization into irreducibles is very easy. What is not so easy is to actually find a factorization for a given polynomial, or even to determine whether the polynomial is itself irreducible. There are not many simple criteria that can be used in this connection.

For polynomials of degree 2 or 3, reducibility is actually equivalent to the existence of a root:

THEOREM 19.8 Let F be a field, and let $f(X) \in F[X]$ have degree 2 or 3. Then $f(X)$ is reducible in $F[X]$ iff $f(X)$ has a root in F.

PROOF. If $f(X)$ has a root $a \in F$, then Theorem 19.3 shows that we can write $f(X) = (X - a)h(X)$, where $\deg(h) = 1$ or 2. Thus $f(X)$ is reducible. Conversely, if $f(X)$ factors as the product of two nonconstant polynomials, then one of these has degree 1 (since $\deg(f) = 2$ or 3). This factor has a root in F, which is also a root of $f(X)$. $\square$

The message of Theorem 19.8 is that we should sometimes use roots to determine reducibility, rather than the other way around.

Examples

1. The polynomial $X^2 + 1$ is irreducible in $\mathbb{R}[X]$, because it has no roots in $\mathbb{R}$. Of course, $X^2 + 1$ factors in $\mathbb{C}[X]$, as $(X - i)(X + i)$. This simple example illustrates the fact that the irreducibility of $f(X) \in F[X]$ depends as much on F as it does on f.

2. Consider the polynomial $f(X) = X^3 - X^2 + 2X - 2$ in $\mathbb{Z}_3[X]$. It is clear that 1 is a root of $f(X)$, and we find by long division that

$$f(X) = (X - 1)(X^2 + 2).$$

Since 1 is again a root of $X^2 + 2$, we obtain

$$f(X) = (X - 1)(X - 1)(X + 1).$$

We could also have factored X^2+2 by observing that $2=-1$ in $\mathbb{Z}_3$, so $X^2+2=X^2-1=(X-1)(X+1)$.

On the other hand, the polynomial $g(X)=X^3+2X+1$ has no roots in $\mathbb{Z}_3$, since $g(0)=g(1)=g(2)=1$. Thus $g(X)$ is irreducible in $\mathbb{Z}_3[X]$.

3. Consider the polynomial $f(X)=X^3+7X^2+3X+1$ in $\mathbb{Q}[X]$. By Exercise 19.1, the only possible roots of $f(X)$ in $\mathbb{Q}$ are ± 1. Since neither of these works, $f(X)$ is irreducible in $\mathbb{Q}[X]$.

4. It is very important to remember that Theorem 19.8 talks only about polynomials of degree 2 or 3. For instance, the fourth-degree polynomial $(X^2+1)(X^2+1)$ is manifestly reducible in $\mathbb{R}[X]$, but it has no roots in $\mathbb{R}$.

In most cases the information provided by Theorem 19.8 is grossly inadequate, and we have to use other methods to try and determine irreducibility. We will now focus our attention on developing some additional techniques that can be used in dealing with the important special case of irreducibility in $\mathbb{Q}[X]$.

Actually there is a method, due to Leopold Kronecker (1823–1891), that will determine the reducibility or irreducibility of any polynomial in $\mathbb{Q}[X]$ in a finite number of steps. Unfortunately, the use of Kronecker's method involves a great deal of tedious calculation ("finite" doesn't mean "small"). The drudgery can be avoided by using a computer, but rather than discuss Kronecker's method we will pursue a couple of simpler criteria that sometimes enable us to see at a glance that polynomials in $\mathbb{Q}[X]$ are irreducible.

We can confine our attention to polynomials with integer coefficients, because if

$$f(X)=\frac{a_0}{b_0}+\frac{a_1}{b_1}X+\cdots+\frac{a_n}{b_n}X^n\in\mathbb{Q}[X],$$

then the irreducibility of $f(X)$ is equivalent to the irreducibility of $g(X)$, where $g(X)\in\mathbb{Z}[X]$ is obtained by multiplying $f(X)$ by the product of the b_i's.

The results we are after depend on the observation that if $f(X)\in\mathbb{Z}[X]$ and we can write

$$f(X)=g(X)h(X)$$

for nonconstant $g,h\in\mathbb{Q}[X]$, then $f(X)$ can already be written as the product of two nonconstant polynomials in $\mathbb{Z}[X]$. To verify this observation, we write $g(X)=(1/c)[g^*(X)]$, $h(X)=(1/d)[h^*(X)]$, with $c,d\in\mathbb{Z}$ and $g^*,h^*\in\mathbb{Z}[X]$; we want to show that we can pull enough common factors out from the coefficients of g^* and the coefficients of h^* to cancel out $1/c$ and $1/d$.

We thus define the **content** of a nonzero polynomial in $\mathbb{Z}[X]$ to be the g.c.d. of its coefficients; we call a polynomial **primitive** if its content is 1. The essential fact about these concepts is

LEMMA 19.9 (Gauss' Lemma) The product of two primitive polynomials is primitive.

PROOF. Suppose

$$g_1(X) = a_0 + a_1 X + \cdots + a_n X^n \qquad \text{and} \qquad h_1(X) = b_0 + b_1 X + \cdots + b_m X^m$$

are two primitive polynomials in $\mathbb{Z}[X]$. We must show that if p is any prime, then there is some coefficient of $g_1(X)h_1(X)$ that is not divisible by p. Let a_k be the highest (i.e., rightmost) coefficient of g_1 that is not divisible by p (there is one since g_1 is primitive), and let b_l be the highest coefficient of h_1 that is not divisible by p. The coefficient of X^{k+l} in $g_1(X)h_1(X)$ is

$$c_{k+l} = a_k b_l + \text{terms of the form } a_i b_j, \quad \text{where either } i > k \text{ or } j > l.$$

Since p does not divide $a_k b_l$ but does divide all the other terms, p does not divide c_{k+l}, and Gauss' Lemma is proved. $\square$

We put it to work:

LEMMA 19.10 Let $f(X) \in \mathbb{Z}[X]$. If $f(X)$ can be written as the product of two nonconstant polynomials in $\mathbb{Q}[X]$, then $f(X)$ can be written as the product of two nonconstant polynomials in $\mathbb{Z}[X]$.

PROOF. As in the discussion preceding the statement of Gauss' Lemma, suppose $g, h \in \mathbb{Q}[X]$ are nonconstant polynomials such that

$$f(X) = g(X)h(X),$$

and write $g(X) = (1/c)g^*(X)$, $h(X) = (1/d)h^*(X)$, with $g^*, h^* \in \mathbb{Z}[X]$. We can now write

$$g^*(X) = \text{content}(g^*) \cdot g_1(X) \quad \text{and} \quad h^*(X) = \text{content}(h^*) \cdot h_1(X),$$

with g_1 and h_1 primitive. Therefore,

$$f(X) = \frac{\text{content}(g^*)\text{content}(h^*)}{cd} g_1(X)h_1(X),$$

and since g_1 and h_1 have the same degrees as g and h, we will be done if we can show that the constant factor on the right-hand side is in $\mathbb{Z}$. Now

$$cd \cdot f(X) = \text{content}(g^*)\text{content}(h^*)g_1(X)h_1(X),$$

so, taking contents on both sides,

$$|cd| \cdot \text{content}(f) = \text{content}(g^*)\text{content}(h^*)\text{content}(g_1 h_1).$$

By Gauss' Lemma, content $(g_1 h_1) = 1$, so we obtain

$$\text{content}(f) = \frac{\text{content}(g^*)\text{content}(h^*)}{|cd|}.$$

Since content$(f) \in \mathbb{Z}$, we have what we want. $\square$

We now reap the first benefit of the preceding discussion by stating a criterion for irreducibility in $\mathbb{Q}[X]$ due to Ferdinand Eisenstein (1823–1852).

Eisenstein was a prize pupil of Gauss, and his result is very much in the spirit of Gauss' Lemma.

THEOREM 19.11 (The Eisenstein Criterion) Let $f(X) = a_0 + a_1 X + \cdots + a_n X^n \in \mathbb{Z}[X]$, and suppose p is a prime such that

$$p \mid a_0, p \mid a_1, \ldots, p \mid a_{n-1}, p \nmid a_n, \quad \text{and} \quad p^2 \nmid a_0.$$

Then $f(X)$ is irreducible in $\mathbb{Q}[X]$.

PROOF. If not, then by Lemma 19.10 there are nonconstant $g(X), h(X) \in \mathbb{Z}[X]$ such that $f(X) = g(X)h(X)$. Say

$$g(X) = b_0 + b_1 X + \cdots + b_m X^m, \qquad h(X) = c_0 + c_1 X + \cdots + c_k X^k,$$
$$\text{with } 1 \leqslant m, k < n.$$

Then $b_0 c_0 = a_0$, so since $p \mid a_0$ but $p^2 \nmid a_0$, exactly one of b_0, c_0 is divisible by p. Say $p \mid b_0, p \nmid c_0$. There is some b_i that is not divisible by p; in fact, since $b_m c_k = a_n, p \nmid b_m$. Let b_r, $1 \leqslant r \leqslant m$, be the first b_i from the left that is not divisible by p. Consider

$$a_r = b_0 c_r + b_1 c_{r-1} + \cdots + b_r c_0.$$

The last term in this sum is not divisible by p, since $p \nmid b_r$ and $p \nmid c_0$. But every other term *is* divisible by p, by the choice of b_r. Thus $p \nmid a_r$. Since $r \leqslant m < n$, this contradicts the given assumptions about p. $\square$

Examples

1. The polynomial $2X^5 + 9X^4 + 3X^2 + 15X + 12$ is irreducible in $\mathbb{Q}[X]$, because the prime 3 divides 9, 3, 15, and 12, but not 2, and $3^2 \nmid 12$.

2. The polynomial $X^2 - 2$ is irreducible in $\mathbb{Q}[X]$, since $2 \mid -2$, $2 \nmid 1$, and $2^2 \nmid -2$. In particular, $X^2 - 2$ has no root in $\mathbb{Q}$, so we have proved that $\sqrt{2}$ is not a rational number.

3. Eisenstein's Criterion does not apply to the polynomial $X^4 + 1$. This does not mean that $X^4 + 1$ is reducible in $\mathbb{Q}[X]$, however, because Eisenstein's Criterion is only a sufficient condition for irreducibility. It isn't a necessary condition, and in the present case, $X^4 + 1$ is in fact irreducible in $\mathbb{Q}[X]$. We can see this by supplementing Eisenstein's Criterion with a simple trick. Namely, if $X^4 + 1 = f(X)g(X)$ in $\mathbb{Q}[X]$, then

$$(X+1)^4 + 1 = f(X+1)g(X+1).$$

That is, the factorization remains if we replace X by $X+1$. Thus the reducibility of $X^4 + 1$ would entail the reducibility of $(X+1)^4 + 1$. But

$$(X+1)^4 + 1 = X^4 + 4X^3 + 6X^2 + 4X + 2,$$

to which Eisenstein's Criterion *does* apply, with $p=2$. Hence X^4+1 is irreducible in $\mathbb{Q}[X]$.

4. Continuing the last example, note that in $\mathbb{R}[X]$ we have

$$X^4+1=(X^2-\sqrt{2}X+1)(X^2+\sqrt{2}X+1).$$

Thus X^4+1 is *reducible* in $\mathbb{R}[X]$, and we see again that the irreducibility of a given polynomial depends on which field we consider the coefficients to lie in.

Our final criterion for irreducibility attempts to answer questions about polynomials over $\mathbb{Q}$ by looking at polynomials over the finite field $\mathbb{Z}_p$, where p is a prime.

THEOREM 19.12 Let p be a prime, and for $m \in \mathbb{Z}$, let $\bar{m}$ denote the remainder of $m \pmod{p}$. Let

$$f(X)=a_0+a_1X+\cdots+a_nX^n$$

be a nonconstant polynomial in $\mathbb{Z}[X]$, and let

$$\bar{f}(X)=\bar{a}_0+\bar{a}_1X+\cdots+\bar{a}_nX^n \in \mathbb{Z}_p[X]$$

be the polynomial obtained by reducing all the coefficients of $f(X)$ mod p. Then if $\bar{f}(X)$ is irreducible in $\mathbb{Z}_p[X]$ and $\deg(\bar{f})=\deg(f)$, $f(X)$ is irreducible in $\mathbb{Q}[X]$.

PROOF. If $f(X)$ is reducible in $\mathbb{Q}[X]$, then by Lemma 19.10 we can write $f(X)=g(X)h(X)$, with $g,h \in \mathbb{Z}[X]$ and $\deg(g) \geq 1$, $\deg(h) \geq 1$. We thus obtain $\bar{f}(X)=\bar{g}(X)\bar{h}(X)$ (see Exercise 19.13), and since $\deg(\bar{f})=\deg(f)$ we must have $\deg(\bar{g})=\deg(g) \geq 1$, $\deg(\bar{h})=\deg(h) \geq 1$. Thus $\bar{f}(X)$ is reducible in $\mathbb{Z}_p[X]$, contrary to our assumption. $\square$

Example The polynomial $X^3+2X+20$ is irreducible in $\mathbb{Q}[X]$, because if we reduce its coefficients mod 3, we get $X^3+2X+2 \in \mathbb{Z}_3[X]$, which has the same degree and is irreducible by Theorem 19.8. Note that if we reduce the coefficients mod 2, we get $X^3 \in \mathbb{Z}_2[X]$, which is reducible. Thus if we reduce mod some prime and get a reducible polynomial, it does not follow that the original polynomial is reducible.

EXERCISES

19.1 Let $f(X)=a_0+a_1X+\cdots+a_rX^r \in \mathbb{Z}[X]$. Suppose $m/n \in \mathbb{Q}$, with $(m,n)=1$. Show that if m/n is a root of $f(X)$, then $m|a_0$ and $n|a_r$.

19.2 Determine which of the following are irreducible in $\mathbb{Q}[X]$.

  a) X^3+X+36

  b) $2X^3-8X^2-6X+20$

c) $2X^4 + 3X^3 + 15X + 6$

d) $X^4 + 2X^3 + X^2 + X + 1$

e) $X^5 + 14X^2 + 4X + 6$

f) $X^4 + X^3 + X^2 + X + 1$ [This is $(X^5 - 1)/(X - 1)$. Recall how we dealt with $X^4 + 1$.]

19.3 Write each polynomial as a product of irreducible polynomials over the given field.

a) $2X^3 + X^2 + 2$, over $\mathbb{Z}_3$

b) $X^3 + 3X^2 + X + 4$, over $\mathbb{Z}_5$

c) $X^2 + 5$, over $\mathbb{Z}_7$

d) $X^4 + X^3 + 2X^2 + X + 2$, over $\mathbb{Z}_3$

e) $X^5 + X^2 - X - 1$, over $\mathbb{Z}_2$

19.4 Suppose that $f(X) \in \mathbb{R}[X]$ and $\alpha = a + bi \in \mathbb{C}$ is a root of $f(X)$. Show that the complex conjugate $\bar{\alpha} = a - bi$ of α is also a root of $f(X)$.

19.5 The Fundamental Theorem of Algebra asserts that every nonconstant polynomial in $\mathbb{C}[X]$ has a factorization in $\mathbb{C}[X]$ of the form

$$f(X) = c(X - c_1)(X - c_2) \cdots (X - c_n), \qquad n \geqslant 1.$$

Assuming this, show that every nonconstant $f(X) \in \mathbb{R}[X]$ can be factored in $\mathbb{R}[X]$ as a product of irreducible polynomials of degree at most 2. (Use the result of Exercise 19.4.)

19.6 Let $f(X) = aX^2 + bX + c \in \mathbb{R}[X]$, $a \neq 0$. Show that $f(X)$ is irreducible in $\mathbb{R}[X]$ iff $b^2 - 4ac < 0$.

19.7 Let $a \in \mathbb{Z}^+$. Show that $X^4 + a$ is reducible in $\mathbb{Q}[X]$ iff $a = 4b^4$ for some integer b.

19.8 Let $f(X), g(X)$ be nonzero polynomials in $\mathbb{Z}[X]$. Show that

$$\text{content}(fg) = \text{content}(f) \cdot \text{content}(g).$$

19.9 Show that the polynomials $q(X)$ and $r(X)$ in Theorem 19.2 are uniquely determined by $f(X)$ and $g(X)$.

19.10 Let F be a field.

a) What are the units in $F[X]$?

b) Show that if $c \in F$, $c \neq 0$, then for any $f(X) \in F[X]$, $f(X)$ and $cf(X)$ generate the same principal ideal in $F[X]$.

19.11 Let R be a ring. Verify the following for $R[X]$:

a) the left distributive law;

b) associativity of multiplication.

19.12 Let R be a commutative ring, $r \in R$, $f(X), g(X) \in R[X]$. Let $h(X) = f(X) + g(X)$ and $k(X) = f(X)g(X)$. Show that

$$h(r) = f(r) + g(r) \qquad \text{and} \qquad k(r) = f(r)g(r).$$

Thus the mapping $\varphi_r: R[X] \to R$ given by $\varphi_r(f(X)) = f(r)$ is a homomorphism. It is called an **evaluation homomorphism**.

19.13 Let R and S be rings, and let $\varphi: R \to S$ be a homomorphism. For $f(X) \in R[X]$, let $\bar{f}(X)$ denote the polynomial in $S[X]$ obtained by replacing each coefficient a_i of f by $\varphi(a_i)$. Show that the mapping $R[X] \to S[X]$ given by $f \to \bar{f}$ is a homomorphism.

19.14 Give another proof of Eisenstein's Criterion, by considering the homomorphism $\mathbb{Z}[X] \to \mathbb{Z}_p[X]$ obtained by reducing all the coefficients of polynomials in $\mathbb{Z}[X]$ mod p.

19.15 Let F be a field, let $b_1, \ldots, b_{n+1}$ be $n+1$ distinct elements of F, and let $c_1, \ldots, c_{n+1}$ be $n+1$ elements of F (not necessarily distinct).

   a) Find a polynomial $f(X) \in F[X]$ such that $f(X) = 0$ or $\deg(f) \leqslant n$, and $f(b_i) = c_i$, $1 \leqslant i \leqslant n+1$.

   b) Show that there is only one such polynomial.

The result of this exercise is often referred to as the **Lagrange Interpolation Theorem**.

19.16 a) Let D be a domain, $f(X) \in D[X]$, $a \in D$. Show that $f(a) = 0$ iff $X - a$ divides $f(X)$ in $D[X]$.

   b) Show that if $f(X)$ has degree n, then f has at most n roots in D.

   c) Show that if D is infinite and two polynomials f, g in $D[X]$ induce the same function on D, then $f = g$.

19.17 Let F be a field. For $f(X) = a_0 + a_1 X + \cdots + a_n X^n \in F[X]$, define the *formal derivative* $f'(X)$ by

$$f'(X) = a_1 + 2a_2 X + 3a_3 X^2 + \cdots + na_n X^{n-1}.$$

   a) Show that if $f, g \in F[X]$ and $h(X) = f(X) + g(X)$, then $h'(X) = f'(X) + g'(X)$.

   b) Show that if $k(X) = f(X)g(X)$ then $k'(X) = f(X)g'(X) + f'(X)g(X)$.

   c) Show that if n is a positive integer, then the formal derivative of $[f(X)]^n$ is $n[f(X)]^{n-1} \cdot f'(X)$.

19.18 Let F be a field, $f(X) \in F[X]$. If $a \in F$ is a root of f, then we can write $f(X) = (X-a)^m g(X)$, where $m \geqslant 1$ and $X - a$ does not divide $g(X)$.

   a) Show that m is uniquely determined by f and a. That is, if we also have $f(X) = (X-a)^r h(X)$, where $X - a$ does not divide $h(X)$, then $r = m$.

   b) If $m \geqslant 2$, we say that a is a *multiple root* of f. Show that a is a multiple root of f iff $f'(a) = 0$. (See Exercise 19.17.)

19.19 Let F be a finite field with q elements.

   a) Show that $a^{q-1} = 1$ for every $a \neq 0$ in F.

   b) Let $f(X) \in F[X]$. Show that there exists a polynomial $f^*(X) \in F[X]$ such that either $f^* = 0$ or $\deg(f^*) < q$, and f^* induces the same function on F as f does.

c) Show that if two polynomials f and g, each of degree $<q$, induce the same function on F, then $f = g$.

19.20 *Kronecker's method.* Let $f(X) \in \mathbb{Z}[X]$ be nonconstant. If f is reducible in $\mathbb{Q}[X]$, then there exist nonconstant $g(X), h(X)$ in $\mathbb{Z}[X]$ such that $f(X) = g(X)h(X)$. Thus for $m \in \mathbb{Z}$, $g(m) | f(m)$. Use this fact together with the result of Exercise 19.15 to describe a method for deciding in a finite number of steps whether $f(X)$ is irreducible over $\mathbb{Q}$.

FROM POLYNOMIALS TO FIELDS

In the preceding section we consolidated our position with respect to some old ideas. We now break new ground, by indicating how polynomial rings can be used to construct interesting new fields.

Our plan is to take the polynomial ring $F[X]$ over a known field F and factor it by a maximal ideal, thus obtaining a field. To see what kinds of fields we can get this way, we first have to ask ourselves what the maximal ideals of $F[X]$ are.

Better yet, what are the ideals?

THEOREM 20.1 If F is a field, every ideal of $F[X]$ is principal.

In other words, if I is any ideal of $F[X]$, then there is some $f(X) \in I$ such that $I = \{ f(X)g(X) | g(X) \in F[X] \}$. According to our usual aR notation for principal ideals, we should denote this ideal by $f(X)F[X]$. However, we will use $(f(X))$ instead, because it is less cumbersome.

PROOF OF THE THEOREM. Let I be an ideal of $F[X]$. If I is the trivial ideal, then $I = (0)$. Otherwise, I contains some nonzero elements, and it is clear that if I is to be $(f(X))$ then, by the degree rule, every nonzero element of I must have degree $\geqslant \deg(f)$.

So let $S = \{ \deg(g) | g \in I \text{ and } g \neq 0 \}$. Since S is a nonempty set of non-negative integers, it has a smallest element, n. Choose $f(X)$ in I such that $\deg(f) = n$. It is clear that $(f(X)) \subseteq I$, and to establish the reverse inclusion we use the division algorithm for $F[X]$.

Let $g(X) \in I$, and write $g(X) = f(X)q(X) + r(X)$, where either $r(X) = 0$ or $\deg(r) < \deg(f)$. Since $g(X) \in I$ and $f(X)q(X) \in I$, we have

$$r(X) = g(X) - f(X)q(X) \in I.$$

Thus $\deg(r) < \deg(f)$ is impossible, by the choice of $f(X)$, and we conclude that $r(X) = 0$. Therefore $g(X) = f(X)q(X) \in (f(X))$, and $I \subseteq (f(X))$, as desired. $\square$

If you think back to how we showed that every ideal of $\mathbb{Z}$ is principal, you'll recall that it essentially came down to an application of the division algorithm for $\mathbb{Z}$. Thus the parallel between the argument for $F[X]$ and the argument for $\mathbb{Z}$ is inescapable.

THEOREM 20.2 Let F be a field. The ideal $(f(X))$ is maximal in $F[X]$ iff $f(X)$ is an irreducible polynomial.

PROOF. First suppose that $f(X)$ is irreducible. To show that $I = (f(X))$ is maximal, we want to show that if J is an ideal of $F[X]$ such that $I \subseteq J$, then either $J = I$ or $J = F[X]$. By Theorem 20.1, we know that $J = (g(X))$ for some $g(X) \in F[X]$, and since $(f(X)) \subseteq (g(X))$ we have $f(X) = g(X)h(X)$ for some $h(X)$. But f is irreducible, so either g or h must be a nonzero constant polynomial. If g is constant, then $J = F[X]$, and if h is constant, then $J = I$.

Conversely, assume that $(f(X))$ is maximal. Then $f(X)$ is nonconstant, since the ideal generated by a constant polynomial is either $\{0\}$ or $F[X]$, and neither of these is maximal. To see that $f(X)$ must be irreducible, suppose that $f(X) = g(X)h(X)$ for two nonconstant polynomials $g(X)$ and $h(X)$, each of degree less than $\deg(f)$. Then neither g nor h can be in $(f(X))$, although their product is. This means that $(f(X))$ is not prime, and therefore not maximal, a contradiction. $\square$

By way of analogy, recall that the maximal ideals of $\mathbb{Z}$ are the ideals $p\mathbb{Z}$, where p is prime.

Example 1 The polynomial $f(X) = X^2 + X + 1$ is irreducible in $\mathbb{Z}_2[X]$, since it has no roots in $\mathbb{Z}_2$. Thus $(f(X))$ is a maximal ideal in $\mathbb{Z}_2[X]$, and $\mathbb{Z}_2[X]/(f(X))$ is a field. Let us denote it by K.

To see what K looks like, notice that the coset $\overline{g(X)}$ determined by any $g(X) \in \mathbb{Z}_2[X]$ is the same as that determined by some $r(X)$ which is either 0 or has degree $\leqslant 1$. This follows from the fact that we can write $g(X) = f(X)q(X) + r(X)$ for some such $r(X)$, so g and r differ by an element of $(f(X))$. Thus the elements of K are

$$\bar{0}, \quad \bar{1}, \quad \bar{X}, \quad \text{and} \quad \overline{X+1},$$

the indicated elements being distinct since no two of 0, 1, X, and $X+1$ differ by an element of $(f(X))$.

Thus K is a field with four elements in it! This is the first finite field we have seen, other than the prime fields $\mathbb{Z}_p$.

Actually, there is something else fascinating about K. For in K we have $X^2 + X + 1 = \bar{0}$, that is,

$$\bar{X}^2 + \bar{X} + \bar{1} = \bar{0}.$$

Since the subfield $\{\bar{0}, \bar{1}\}$ of K is isomorphic to $\mathbb{Z}_2$, we can regard $\mathbb{Z}_2$ as a subfield of K, by replacing $\bar{0}$ and $\bar{1}$ by 0 and 1. If we do this, then the above equation reads

$$\bar{X}^2 + \bar{X} + 1 = 0,$$

in other words $f(\bar{X}) = 0$. Thus although f had no root in $\mathbb{Z}_2$, it has acquired one in the extension K! We have actually manufactured a solution to the equation $f(X) = 0$.

Magic, no?

Let's try it again.

Example 2 The polynomial $f(X) = X^2 + 1$ is irreducible in $\mathbb{R}[X]$. Thus $\mathbb{R}[X]/(f(X))$ is a field, which we will call C. By the division algorithm for $\mathbb{R}[X]$, the elements of C are the elements

$$\overline{a + bX} = \bar{a} + \bar{b}\bar{X},$$

for $a, b \in \mathbb{R}$. In C we have $\overline{X^2 + 1} = \bar{0}$, that is, $\bar{X}^2 + \bar{1} = \bar{0}$.

The subset $\{\bar{a} | a \in \mathbb{R}\}$ is a subfield of C isomorphic to $\mathbb{R}$. If we replace each $\bar{a}$ by a, then we have $\mathbb{R} \subseteq C$, and $C = \{a + b\bar{X} | a, b \in \mathbb{R}\}$. The element $\bar{X}$ in C is a root of the polynomial $X^2 + 1$.

Thus $\bar{X}$ is a square root of -1, and it follows that the addition and multiplication in C satisfy

$$(a + b\bar{X}) + (c + d\bar{X}) = (a + c) + (b + d)\bar{X},$$
$$(a + b\bar{X})(c + d\bar{X}) = (ac - bd) + (ad + bc)\bar{X}.$$

Therefore $C \cong \mathbb{C}$, and we have once again constructed $\mathbb{C}$ from $\mathbb{R}$ (compare Exercise 16.5).

The preceding examples are illustrations of a general phenomenon:

THEOREM 20.3 (Kronecker) Let F be a field and let $f(X) \in F[X]$ have degree $\geqslant 1$. Then there is an extension field K of F that contains a root of $f(X)$.

PROOF. We can assume that $f(X)$ is irreducible in $F[X]$, because if it isn't we can work with one of its irreducible factors.

Under this assumption, $F[X]/(f(X))$ is a field, in which $\overline{f(X)} = \bar{0}$. If $f(X) = a_0 + a_1 X + \cdots + a_n X^n$, this means that

$$\bar{a}_0 + \bar{a}_1 \bar{X} + \cdots + \bar{a}_n \bar{X}^n = \bar{0}. \qquad [20.1]$$

We have an embedding $\varphi : F \to F[X]/(f(X))$, given by $\varphi(a) = \bar{a}$, for every $a \in F$. φ is clearly a homomorphism, and it is one-to-one since if $a_1, a_2 \in F$ and

$\bar{a}_1 = \bar{a}_2$, then $f(X)$ divides $a_1 - a_2$ in $F[X]$. This means that $a_1 - a_2 = 0$, since $\deg(f) \geqslant 1$, so $a_1 = a_2$.

Thus $\{\bar{a} \mid a \in F\}$ is a subfield of $F[X]/(f(X))$, isomorphic to F. If we replace each element $\bar{a}$ by a, we obtain a field K that extends F, in which $a_0 + a_1\bar{X} + \cdots + a_n\bar{X}^n = 0$, by [20.1]. $\square$

Theorem 20.3 is the key to an in-depth analysis of the roots of polynomials, because it tells us we can always get our hands on a complete set of roots to work with:

COROLLARY 20.4 Let $f(X) \in F[X]$ have degree $n \geqslant 1$. Then there is a field $K \supseteq F$ such that in $K[X]$ we can write $f(X) = a(X - c_1)(X - c_2) \cdots (X - c_n)$.

PROOF. By induction on n. For $n = 1$ we have $f(X) = aX + b$, with $a, b \in F$, $a \neq 0$, so we can write $f(X) = a(X - (-b/a))$ in $F[X]$. Now suppose $\deg(f) = m$ and the result is proved for polynomials of degree $m - 1$. By Theorem 20.3, we can let $K_1 \supseteq F$ be a field containing a root c_1 of $f(X)$. In $K_1[X]$, we can write $f(X) = (X - c_1)g(X)$, where $g(X)$ has degree $m - 1$ and is therefore subject to the inductive hypothesis. Let $K \supseteq K_1$ be such that in $K[X]$ we have

$$g(X) = a(X - c_2) \cdots (X - c_n).$$

Then K is an extension of F, and we have

$$f(X) = a(X - c_1) \cdots (X - c_n)$$

in $K[X]$. $\square$

Of course, some of $c_1, \ldots, c_n$ may already lie in F, and $c_1, \ldots, c_n$ need not be distinct.

Example Let $f(X) = X^4 - X^3 - X + 1 \in Z_5[X]$. Then 1 is a root of $f(X)$, and long division yields

$$f(X) = (X - 1)(X^3 - 1)$$

in $Z_5[X]$. Since 1 is again a root of $X^3 - 1$, we obtain

$$f(X) = (X - 1)(X - 1)(X^2 + X + 1).$$

The factor $X^2 + X + 1$ has no roots in Z_5, but we know there is a root c in some extension K of Z_5. We divide $X^2 + X + 1$ by $X - c$ in $K[X]$:

$$
\begin{array}{r}
X + (c+1) \\
X - c \overline{\smash{\big)}\ X^2 + X + 1 } \\
\underline{X^2 - cX } \\
(c+1)X + 1 \\
\underline{(c+1)X - c(c+1)} \\
c(c+1) + 1 = c^2 + c + 1 = 0.
\end{array}
$$

Thus, in $K[X]$,

$$f(X)=(X-1)(X-1)(X-c)(X-(-c-1)).$$

EXERCISES

20.1 Let p be a prime. Show that $\mathbb{Z}_p[X]/(X^2+1)$ is a field iff the equation $x^2 \equiv -1$ has no solution (mod p).

20.2 Is $\mathbb{Q}[X]/((X-1)^2)$ a domain?

20.3 Is $\mathbb{Q}[X]/(X^3+2X+2)$ a field? How about $\mathbb{R}[X]/(X^3+2X+2)$?

20.4 Let $K=\{0,1,\overline{X},\overline{X}+1\}$ be the four-element field constructed in Example 1 on pp. 201–202. Write X^2+X+1 as a product of factors of degree 1 in $K[X]$.

20.5 The elements of the field K in Exercise 20.4 can each be written uniquely in the form $a+b\overline{X}$, with $a,b \in \mathbb{Z}_2$. Find a general rule for writing the product $(a+b\overline{X})(c+d\overline{X})$ in this form.

20.6 a) Let $f(X)$ be irreducible in $F[X]$, and let K be the field obtained from $F[X]/(f(X))$ by replacing $\bar{a}$ by a, for each $a \in F$. Show that if $\deg(f)=n$, then every element of K has a unique representation in the form

$$a_0+a_1\overline{X}+\cdots+a_{n-1}\overline{X}^{n-1},$$

with $a_i \in F$.

b) Show that if we start with $F=\mathbb{Z}_p$ and $\deg(f)=n$, then the field K in part (a) has p^n elements.

20.7 Use the result of Exercise 20.6 to construct a field with m elements, for $m=$

a) 8;

b) 9;

c) 27;

d) 25;

e) 125.

20.8 Let F be a field. What are the prime ideals in $F[X]$?

20.9 Let $n \geqslant 1$, let p be a prime, and let $f(X)=X^{p^n}-X \in \mathbb{Z}_p[X]$. By Corollary 20.4, let K be an extension of $\mathbb{Z}_p$ such that in $K[X]$ we have

$$f(X)=(X-c_1)\cdots(X-c_{p^n}).$$

a) Show that in this case $c_1,\ldots,c_{p^n}$ are all distinct. (Use Exercise 19.18.)

b) Show that $\{c_1,\ldots,c_{p^n}\}$ is a subfield of K, and hence that there exists a field with p^n elements.

UNIQUE FACTORIZATION DOMAINS

We have called a positive integer p a prime if $p \neq 1$ and the only positive divisors of p are 1 and p itself. This notion of "prime" generalizes in a natural way to allow negative primes: an integer n is called **prime** if $n \neq 0$, $n \neq \pm 1$, and the only divisors of n are $\pm 1, \pm n$. Thus $2, -2, 3, -3, 5, -5, 7, -7, 11, -11 \ldots$ are all primes in $\mathbb{Z}$.

In these terms, the Fundamental Theorem of Arithmetic may be taken as the statement that every integer n which is neither zero nor ± 1 can be factored into a product of primes, and that if

$$n = p_1 p_2 \cdots p_r \quad \text{and} \quad n = q_1 q_2 \cdots q_s$$

are two such factorizations, then $r = s$ and, after rearranging the q_i's if necessary, we have $p_i = \pm q_i$. For example, we have $-60 = (-2)(2)(3)(5) = (-3)(2)(-5)(-2)$, and if we rearrange the second factorization it becomes $(-2)(2)(-3)(-5)$.

The Fundamental Theorem of Arithmetic seems obvious to most of us, because we all grew up with it in school. For this reason, it is easy to get lulled into thinking that the theorem doesn't require any proof. One purpose of this section is to convince you that it does require proof by showing you some integral domains for which the analogue of the Fundamental Theorem is false. With such examples in hand, it becomes interesting to try to enunciate some conditions on a domain which will guarantee that the analogue of the Fundamental Theorem does hold.

The necessity for such an investigation is underscored by the fact that some notable mathematicians have fallen into the trap of assuming that the Fundamental Theorem holds in cases where in fact it does not. Efforts to salvage the results which they based on these faulty assumptions have been largely responsible for the development of ring theory as we know it today.

Before we recount a specific instance, let us establish some vocabulary, so that we can say things a bit more precisely.

Let D be a domain. (We concentrate on domains because domains are the natural generalization of $\mathbb{Z}$.) An element $d \in D$ is called **irreducible** if d is neither 0_D nor a unit, and whenever $a, b \in D$ and $d = ab$, then either a or b is a unit.

Examples

1. Since the units in $\mathbb{Z}$ are ± 1, the irreducible elements of $\mathbb{Z}$ are precisely the primes.

2. If F is a field, the units in $F[X]$ are the nonzero elements of F. Thus the irreducible elements of $F[X]$ are the nonconstant polynomials that are irreducible according to the meaning of the word in Section 19.

3. We claim that a nonconstant polynomial $f(X) \in \mathbb{Z}[X]$ is irreducible in $\mathbb{Z}[X]$ iff $f(X)$ is primitive and cannot be written as the product of two nonconstant polynomials in $\mathbb{Z}[X]$.

First of all, if f is irreducible then f must be primitive, for otherwise we could write $f(X) = p \cdot g(X)$ for some prime p and nonconstant $g(X)$, and neither p nor $g(X)$ is a unit in $\mathbb{Z}[X]$. Likewise, f cannot be written as the product of two nonconstant polynomials.

Conversely, suppose that f has the two indicated properties. Then if $f(X) = g(X)h(X)$ in $\mathbb{Z}[X]$, either g or h must be a constant; since f is primitive, this constant must be ± 1, so it is a unit in $\mathbb{Z}[X]$. Thus f is irreducible.

It is easy to see that a constant is irreducible in $\mathbb{Z}[X]$ iff it is a prime in $\mathbb{Z}$, and therefore the irreducible elements of $\mathbb{Z}[X]$ are the primes of $\mathbb{Z}$ and the nonconstant irreducibles described above.

In light of Example 2, our use of the term "irreducible element" seems very natural; but in terms of Example 1, "prime element" might seem more appropriate. In general, the word "prime" is used in connection with a concept that is sometimes stronger than irreducibility. If D is a domain, and $x, y \in D$, then we say that x **divides** y, and we write $x \mid y$, if there is some $z \in D$ such that $y = xz$. An element $d \in D$ is called **prime** if d is neither 0_D nor a unit and whenever $a, b \in D$ and $d \mid ab$, then $d \mid a$ or $d \mid b$.

Prime elements are always irreducible, in any integral domain:

THEOREM 21.1 Let D be a domain, and let $d \in D$ be prime. Then d is irreducible.

PROOF. Suppose $d = ab$; we must show that either a or b is a unit. Since d is prime, we know that d divides either a or b. If $d|a$, then we have $a = dc$ for some c, and so

$$d = (dc)b.$$

Since $d \neq 0$ and D is a domain, we conclude that $1 = cb$ and b is a unit. Similarly, if $d|b$ then a is a unit. $\square$

For $(\mathbb{Z}, +, \cdot)$, the notions of "prime" and "irreducible" coincide (Theorem 4.3) and in general the truth of the implication "irreducible $\Rightarrow$ prime" for a domain D is very intimately connected with the truth of an analogue of the Fundamental Theorem for D (see Exercise 21.9). We will soon see examples of irreducibles that are not prime.

Two elements a and b of a domain D are said to be **associates** if $a = bu$ for some unit u. For example, any element a is an associate of itself, because $a = a \cdot 1$. Similarly, if a and b are associates, then b and a are associates, because $a = bu$ implies $b = au^{-1}$. The relation

$$a R b \quad \text{iff } a \text{ and } b \text{ are associates}$$

is in fact an equivalence relation; transitivity is easily verified.

We now have the words we need in order to say precisely what it means for an analogue of the Fundamental Theorem to hold for D.

DEFINITION Let D be an integral domain. Then D is called a **unique factorization domain** (UFD) if:

i) Every element $d \in D$ that is neither 0 nor a unit can be factored as the product of a finite number of irreducible elements; and

ii) If $d = p_1 p_2 \cdots p_r$ and $d = q_1 q_2 \cdots q_s$ are two such factorizations, then $r = s$, and there is a permutation f of $\{1, 2, \ldots, s\}$ such that p_i and $q_{f(i)}$ are associates, for each $i \in \{1, 2, \ldots, s\}$.

Note that this is a direct generalization of the statement of the Fundamental Theorem that we gave on p. 205, because in $\mathbb{Z}$, two elements a and b are associates iff $a = \pm b$.

Now for a bit of history. We all know that there are triples x, y, z of nonzero integers such that $x^2 + y^2 = z^2$; for instance, $3^2 + 4^2 = 5^2$. In 1637, Pierre de Fermat claimed to have discovered a "truly remarkable proof" that if $n > 2$ then there do *not* exist nonzero integers x, y, z such that $x^n + y^n = z^n$. Fermat never published a proof, and nobody else has ever succeeded in finding one. Because of this, there is a great deal of doubt as to whether Fermat ever had a correct proof. The result is referred to as the **Fermat Conjecture**, or (if you want to be a little kinder to Fermat) **Fermat's Last Theorem**.

The problem is very easily reduced to showing that $x^n + y^n = z^n$ is impossible (for nonzero x,y,z) if n is 4 or an odd prime. (It is fun to do this for yourself, and we won't spoil it by showing you how.) The case $n = 4$ can be disposed of by elementary means, and we know that Fermat did prove this case himself. In any event, the general problem boils down to proving that $x^p + y^p = z^p$ is impossible if p is an odd prime and $xyz \neq 0$.

A proof for $p = 3$ was published in 1770 by Leonhard Euler, the most prolific mathematician of all time. Euler's proof involved using numbers of the form $a + b\sqrt{-3}$, where $a, b \in \mathbb{Z}$ and $\sqrt{-3} = \sqrt{3}i$. At one point in his argument, he made a claim about these numbers which was apparently based on the tacit assumption that they obey unique factorization. His claim was correct, but the tacit assumption behind it was not, and his proof remained incomplete until the missing justification was supplied by Legendre some time later.

A proof for $p = 5$ was given by Legendre, and independently by Dirichlet, around 1825. The case $p = 7$ was handled by Lamé in 1840. The first general —and by far the most significant—attack on the problem was made by E. Kummer in 1843. Kummer's basic idea was to consider numbers of the form

$$a_0 + a_1\zeta_p + a_2\zeta_p^2 + \cdots + a_{p-1}\zeta_p^{p-1},$$

where $a_i \in \mathbb{Z}$ and ζ_p is a complex number $\neq 1$ such that $\zeta_p^p = 1$. (For example, $\zeta_p = \cos(2\pi/p) + i\sin(2\pi/p)$.[†]) These numbers form a subring of $\mathbb{C}$, which we denote by $\mathbb{Z}[\zeta_p]$. Using them, it is possible to factor $x^p + y^p$ completely, and the equation $x^p + y^p = z^p$ becomes

$$(x+y)(x+y\zeta_p)(x+y\zeta_p^2)\cdots(x+y\zeta_p^{p-1}) = z^p.$$

Assuming that $\mathbb{Z}[\zeta_p]$ is a UFD, Kummer used this form of the equation to prove that $x^p + y^p = z^p$ is impossible if $xyz \neq 0$.

Kummer presented his proof to Dirichlet (a more established mathematician), who pointed out that Kummer had neglected to verify the assumption that factorization into irreducibles is unique in $\mathbb{Z}[\zeta_p]$. (Kummer was later to point out a similar flaw in an attempt by Lamé.) In 1847, Cauchy (after having made the same mistake himself) pointed out that factorization is *not* unique in $\mathbb{Z}[\zeta_{23}]$. Thus Fermat's Last Theorem remained unproved.

[†]That $[\cos(2\pi/p) + i\sin(2\pi/p)]^p = 1$ follows from *De Moivre's Theorem*:

$$[\cos\theta + i\sin\theta]^k = \cos k\theta + i\sin k\theta.$$

De Moivre's Theorem can be proved by induction on k, using the fact that

$$[\cos\alpha + i\sin\alpha][\cos\beta + i\sin\beta] = (\cos\alpha\cos\beta - \sin\alpha\sin\beta) + i(\sin\alpha\cos\beta + \cos\alpha\sin\beta)$$
$$= \cos(\alpha+\beta) + i\sin(\alpha+\beta).$$

Undaunted, Kummer set about trying to modify $\mathbb{Z}[\zeta_p]$ so as to restore the uniqueness of factorization. He introduced what he called *ideal numbers*, and the theory he developed was a precursor of the modern theory of ideals. Kummer succeeded in proving Fermat's Last Theorem for certain primes which he called *regular*. Using his ideas, it has since been possible to prove the theorem for all primes < 4001. (In the case where none of x, y, z is divisible by p, it is known, by other methods, to be true for $p < 253,747,889$.) But the general theorem still awaits proof.

In terms of their enduring influence on the development of algebra, the *methods* that Kummer invented in his work on this problem are probably more significant than the progress he made on the problem itself. Attempts by number theorists to exploit the properties of specific systems more inclusive than $\mathbb{Z}$ led naturally to a study of such systems in general, hence to the emergence of an abstract theory of rings.

After we have developed some more information about UFDs, we will discuss in detail the application of another extended number system (the ring of Gaussian integers) to the proof of Fermat's classic "two squares" theorem. But first we want to show you an explicit example of the failure of unique factorization in a fairly simple domain.

In fact, let us use the set of all complex numbers of the form $a + b\sqrt{-3}$, with $a, b \in \mathbb{Z}$, referred to above. If we denote this set by $\mathbb{Z}[\sqrt{-3}]$, then $\mathbb{Z}[\sqrt{-3}]$ is a subring of $\mathbb{C}$ containing 1, hence an integral domain. In $\mathbb{Z}[\sqrt{-3}]$ we have the following two factorizations of the number 4:

$$4 = 2 \cdot 2 = (1 + \sqrt{-3})(1 - \sqrt{-3}).$$

We are going to show that 2, $1 + \sqrt{-3}$, and $1 - \sqrt{-3}$ are all irreducible in $\mathbb{Z}[\sqrt{-3}]$ and that no two of them are associates. This will establish that $\mathbb{Z}[\sqrt{-3}]$ is not a UFD.

For $\alpha = a + b\sqrt{-3}$, define the **norm** $N(\alpha)$ of α to be $N(\alpha) = a^2 + 3b^2$. For every $\alpha \in \mathbb{Z}[\sqrt{-3}]$, $N(\alpha)$ is a nonnegative integer, and $N(\alpha) = 0$ iff $\alpha = 0$. An easy calculation shows that if $\alpha, \beta \in \mathbb{Z}[\sqrt{-3}]$, then

$$N(\alpha\beta) = N(\alpha)N(\beta).$$

Using norms, we can readily determine all the units in $\mathbb{Z}[\sqrt{-3}]$. For if α is a unit, then $\alpha\beta = 1$ for some β, whence

$$N(\alpha)N(\beta) = N(1) = 1,$$

which implies that $N(\alpha) = 1$, since $N(\alpha)$ and $N(\beta)$ are both nonnegative integers. Conversely, if $N(\alpha) = 1$ and, say, $\alpha = a + b\sqrt{-3}$, then $a^2 + 3b^2 = 1$, which implies that $a = \pm 1$ and $b = 0$. Thus $\alpha = \pm 1$, and α is a unit. Thus we see that α is a unit iff $N(\alpha) = 1$, and the only units in $\mathbb{Z}[\sqrt{-3}]$ are ± 1.

Now we claim that 2 is irreducible in $\mathbb{Z}[\sqrt{-3}]$. Clearly, 2 is not 0, and it is not a unit since $2 \neq \pm 1$. We must show that if $2 = \alpha\beta$, then either α or β

must be a unit, that is, either $N(\alpha)$ or $N(\beta)$ must be 1. Now if $2 = \alpha\beta$, then

$$N(2) = N(\alpha)N(\beta)$$
$$4 = N(\alpha)N(\beta),$$

so all we have to show is that neither $N(\alpha)$ nor $N(\beta)$ can be 2. But the norm of any element of $\mathbb{Z}[\sqrt{-3}\,]$ has the form $a^2 + 3b^2$, and it is clear that this never gives us 2.

The same argument shows that both $1 + \sqrt{-3}$ and $1 - \sqrt{-3}$ are irreducible, because their norms are both 4.

No two of our three irreducible elements are associates, because the only units are ± 1. Thus $\mathbb{Z}[\sqrt{-3}\,]$ is not a UFD.

Notice a couple of other interesting things. The equation

$$2\cdot 2 = (1 + \sqrt{-3}\,)(1 - \sqrt{-3}\,)$$

shows that 2 divides the product $(1 + \sqrt{-3}\,)(1 - \sqrt{-3}\,)$, but it obviously does not divide either factor. Thus 2 is irreducible, but not prime. The same can be said for $1 + \sqrt{-3}$ and $1 - \sqrt{-3}$.

Observe that if $\alpha \in \mathbb{Z}[\sqrt{-3}\,]$ and $N(\alpha)$ is a prime integer, then α is irreducible. For if $\alpha = \beta\gamma$, then $N(\alpha) = N(\beta)N(\gamma)$, so one of $N(\beta)$, $N(\gamma)$ is 1. The converse is false; for example, 2 is irreducible, but $N(2) = 4$ is not a prime integer.

By considering norms, it can be shown that every element of $\mathbb{Z}[\sqrt{-3}\,]$ which is neither 0 nor a unit can be factored into a product of irreducibles (Exercise 21.5), and thus it is only the *uniqueness* of factorization that fails, not the existence. On the other hand, it is not true that every element $\neq 0, \pm 1$ can be written as a product of *prime* elements. For instance, 2 is not a prime, and if

$$2 = \alpha_1\alpha_2\cdots\alpha_k, \quad k > 1,$$

then

$$N(2) = 4 = N(\alpha_1)N(\alpha_2)\cdots N(\alpha_k),$$

so some α_i has norm 4 and the others are units, not primes.

We have not yet exhibited a prime in $\mathbb{Z}[\sqrt{-3}\,]$. An example is given in Exercise 21.22.

Under what conditions might it be impossible to factor an element into irreducible factors? Suppose D is a domain, $a \in D$ is neither 0 nor a unit, and a cannot be written as the product of a finite number of irreducible elements. In particular, then, a is not itself irreducible, so we can write $a = a_1 b_1$, where neither a_1 nor b_1 is a unit. Furthermore, our assumptions on a imply that a_1 and b_1 cannot both be written as products of irreducible elements. Say b_1 cannot be so written. We have

$a = a_1 b_1$; a_1, b_1 are not units; b_1 is not a product of irreducible elements.

$$[21.1]$$

Now b_1 has the same properties that a had when we started, so we can write $b_1 = a_2 b_2$, where neither a_2 nor b_2 is a unit, and, say, b_2 cannot be written as a product of irreducible elements. Looking at b_2, we write $b_2 = a_3 b_3$, where neither a_3 nor b_3 is a unit, and b_3 cannot be written as a product of irreducible elements. Clearly we can continue like this indefinitely by induction; we have $b_n = a_{n+1} b_{n+1}$, with a_{n+1}, b_{n+1} not units and b_{n+1} not a product of irreducible elements. We have the following equations:

$$a = a_1 b_1 = a_1 a_2 b_2 = a_1 a_2 a_3 b_3 = a_1 a_2 a_3 a_4 b_4 = \cdots .$$

What conditions on D could possibly rule this out? Think what the recurring situation [21.1] means. If we have $a = a_1 b_1$, then $a \in b_1 D$, the principal ideal of D generated b_1, and this implies that $aD \subseteq b_1 D$, since $b_1 D$ is an ideal. Is $b_1 D \subseteq aD$ too? If so, then in particular $b_1 \in aD$, so $b_1 = ac$ for some c. Then

$$b_1 = ac = a_1 b_1 c.$$

Since $b_1 \neq 0$ and D is a domain, this yields $1 = a_1 c$, and a_1 is a unit. Contradiction!

Thus $aD \subsetneq b_1 D$. Since b_2 was obtained from b_1 just as b_1 was obtained from a, we get $b_1 D \subsetneq b_2 D$ by an identical argument. And, in general, $b_n D \subsetneq b_{n+1} D$, so we have a strictly increasing chain of ideals:

$$aD \subsetneq b_1 D \subsetneq b_2 D \subsetneq b_3 D \subsetneq b_4 D \subsetneq \cdots .$$

The union of this chain, $aD \cup b_1 D \cup b_2 D \cup \cdots$, is easily seen to be an ideal. If this ideal were principal, then the fact that our chain of ideals is strictly increasing would be contradicted. For if

$$aD \cup b_1 D \cup b_2 D \cup b_3 D \cup \cdots = dD,$$

then $d \in b_i D$ for some i, hence $dD \subseteq b_i D$, and the chain would stop at $b_i D$.

Thus one way to ensure that elements factor into irreducibles is to assume that *every* ideal in D is principal. An integral domain with the property that every ideal is principal is called a **principal ideal domain** (PID). For example, $\mathbb{Z}$ is a PID, and so is $F[X]$, for any field F. Our discussion proves

THEOREM 21.2 Let D be a PID, and let $a \in D$. If a is neither 0 nor a unit, then a can be written as the product of finitely many irreducible elements.

This theorem constitutes half of a proof that every PID is a UFD. The other half deals with the uniqueness of the factorization, and for this we take a hint from $\mathbb{Z}$. The key to the uniqueness of factorization for $\mathbb{Z}$ (Exercise 4.28) was the fact that if p were irreducible and p divided a product ab, then p had to divide either a or b. In our current terminology this says that every irreducible element was prime.

We thus rest our hope for unique factorization in PID's on

THEOREM 21.3 Every irreducible element of a PID is prime.

We will establish Theorem 21.3 by proving the following sharper result, which also generalizes Theorem 20.2.

THEOREM 21.4 Let D be a PID, let $d \in D$, and assume that d is neither 0_D nor a unit. Then the following are equivalent:

i) d is prime

ii) d is irreducible

iii) dD is a maximal ideal

iv) dD is a prime ideal.

PROOF. The proof is not much longer than the statement of the theorem; we will show that (i) $\Rightarrow$ (ii) $\Rightarrow$ (iii) $\Rightarrow$ (iv) $\Rightarrow$ (i).

(i) $\Rightarrow$ (ii): See Theorem 21.1.

(ii) $\Rightarrow$ (iii): Since d is not a unit, $1 \notin dD$, so dD is a proper ideal. We must show that if $dD \subsetneq bD$, then $bD = D$. Now if $dD \subsetneq bD$, then in particular $d \in bD$, so we have $d = bc$ for some c. Since d is irreducible, either b or c must be a unit. If c is a unit, then $b = dc^{-1}$ and $b \in dD$, so $bD \subseteq dD$, a contradiction. Thus b is a unit, and $bD = D$.

(iii) $\Rightarrow$ (iv): See Corollary 17.8.

(iv) $\Rightarrow$ (i): Suppose dD is prime, and suppose d divides a product bc. Then $bc \in dD$, so either $b \in dD$ or $c \in dD$, that is, either $d|b$ or $d|c$, as desired. $\square$

THEOREM 21.5 Every PID is a UFD.

PROOF. By Theorem 21.2, it suffices to show that if

$$p_1 p_2 \cdots p_r = q_1 q_2 \cdots q_s \qquad [21.2]$$

for irreducible elements p_i, q_j, then $r = s$ and there is a permutation f of $\{1, 2, \ldots, s\}$ such that p_i and $q_{f(i)}$ are associates for each $i \in \{1, 2, \ldots, s\}$.

Now, from the above equation, $p_1 | q_1 q_2 \cdots q_s$, so by Theorem 21.3 and induction, p_1 divides some q_j. By renaming the q's if necessary, we can assume $j = 1$. Thus $q_1 = ap_1$ for some a, and since q_1 and p_1 are irreducible, this means that a is a unit, so q_1 and p_1 are associates. We have

$$p_1 p_2 \cdots p_r = ap_1 q_2 \cdots q_s,$$

and since we are in a domain, this becomes

$$(a^{-1} p_2) p_3 \cdots p_r = q_2 q_3 \cdots q_s. \qquad [21.3]$$

We now proceed by induction on r. If $r = 1$, then s must be 1, else Equation [21.3] would show that the irreducible element q_2 is a unit, which is nonsense. Thus if $r = 1$, then the original equation [21.2] had one irreducible on each side, and $p_1 = q_1$.

Assuming the result for $r = n$, suppose that $r = n + 1$. Then by [21.3], $s > 1$, and by the inductive hypothesis, $r - 1 = s - 1$ and there is a permutation g of $\{2, \ldots, s\}$ such that p_i and $q_{g(i)}$ are associates for $2 \leqslant i \leqslant s$. (Actually, $a^{-1}p_2$ is an associate of $q_{g(2)}$, but that makes p_2 one, too.) So we have $r = s$, and we can extend g to a permutation of $\{1, 2, \ldots, s\}$ by defining $f(1) = 1$; then p_i and $q_{f(i)}$ are associates for all $i \in \{1, 2, \ldots, s\}$, and we are done. $\square$

Examples

1. Specialized to $\mathbb{Z}$, Theorem 21.5 proves the Fundamental Theorem of Arithmetic.

2. Let F be a field. Applying Theorem 21.5 to $F[X]$, we see that the factorization of polynomials over F into irreducible factors is unique, except for nonzero constant factors and the order in which the factors are written.

3. The converse of Theorem 21.5 is false; there exist UFDs that are not PIDs.

For instance, consider the domain $\mathbb{Z}[X]$, and let I be the ideal consisting of all elements whose constant term is even. I is not principal, since $2 \in I$ and $X \in I$, but there is clearly no element $f(X) \in I$ such that 2 and X are both multiples of $f(X)$. Thus $\mathbb{Z}[X]$ is not a PID.

It is a UFD, however. We will leave the proof that every element factors into irreducibles as an exercise, and concentrate on the uniqueness. So, recalling our description of the irreducibles of $\mathbb{Z}[X]$, suppose that

$$p_1 p_2 \cdots p_r f_1(X) \cdots f_k(X) = q_1 q_2 \cdots q_s g_1(X) \cdots g_l(X),$$

where the p's and q's are primes in $\mathbb{Z}$ and the f's and g's are nonconstant primitive polynomials in $\mathbb{Z}[X]$ that cannot be written as products of nonconstant factors. By Gauss' Lemma and induction, the product of the f's is primitive, and likewise for the product of the g's. It follows (by taking contents) that

$$p_1 p_2 \cdots p_r = \pm q_1 q_2 \cdots q_s,$$

so by unique factorization in $\mathbb{Z}$, $r = s$ and, after a permutation, p_i and q_i are associates. Moreover, since

$$f_1(X) \cdots f_k(X) = \pm g_1(X) \cdots g_l(X)$$

and the f's and g's are irreducible in $\mathbb{Q}[X]$ (by Lemma 19.10), the uniqueness of factorization in $\mathbb{Q}[X]$ implies that $k = l$ and, after a permutation, f_i and g_i are associates in $\mathbb{Q}[X]$. Since f_i and g_i are primitive, this implies that they are associates in $\mathbb{Z}[X]$.

Theorem 21.5 establishes an aesthetically appealing condition which will guarantee that a domain is a UFD. But we must admit that, in practice, it may not be the easiest thing in the world to determine whether or not a given domain is a PID. In terms of the domains that intervene in number theory, for example, one would like to have some practical way to try and see if they really are PIDs (hence UFDs).

Once again we go back to known examples and recall how we showed that $\mathbb{Z}$ and $F[X]$ are PIDs. The argument for $\mathbb{Z}$ amounted to choosing the smallest positive element n in a nontrivial ideal I, and using the division algorithm to show that $I = n\mathbb{Z}$. The argument for $F[X]$ was similar, starting with an element of smallest degree. It seems clear that the same kind of reasoning should work again, any time we have a suitable way of associating nonnegative integers ("degrees") to the elements of the domain in question.

DEFINITION Let D be a domain. Suppose there exists a function $v: D - \{0_D\} \to \mathbb{Z}^+ \cup \{0\}$ such that

i) $v(a) \leqslant v(ab)$ for all nonzero $a, b \in D$; and

ii) for any $a, b \in D$, with $b \neq 0_D$, there exist q and r in D such that $a = qb + r$ and either $r = 0_D$ or $v(r) < v(b)$.

Then we say that D is a **Euclidean domain**.

The nomenclature arises from the fact that in a Euclidean domain, the function v enables us to develop an analogue of the Euclidean algorithm for determining g.c.d.'s (see Exercises 21.11, 21.14, and 21.18). What we want to know about Euclidean domains at the moment is

THEOREM 21.6 Any Euclidean domain is a PID (hence a UFD).

PROOF. Let I be an ideal of the Euclidean domain D. If $I = \{0\}$, then clearly I is principal. If I is not trivial, then the set $\{v(d) | d \in I$ and $d \neq 0\}$ is a nonempty set of nonnegative integers, hence has a least element, n. Let $d \in I$ be such that $v(d) = n$; we assert that $I = Dd$. For if $x \in I$ we can find $q, r \in D$ such that $x = qd + r$ and either $r = 0_D$ or $v(r) < v(d)$. But the fact that both x and d are in I means that $r = x - qd$ is in I too, so $v(r) < v(d)$ is impossible by the choice of d. Hence $r = 0_D$, and $x = qd \in Dd$. Thus $I \subseteq Dd$, and since trivially $Dd \subseteq I$, we are done. $\square$

Examples

1. $\mathbb{Z}$ is a Euclidean domain, with $v(n) = |n|$ for every $n \in \mathbb{Z} - \{0\}$. Notice that we could regard v as being defined on all of $\mathbb{Z}$, here, with $v(0) = |0| = 0$. However, the general theory of Euclidean domains works out neatest if we regard $v(0)$ as being undefined.

2. Let F be any field, and define $v(a)=0$ for $a \neq 0_F$. This makes F into a Euclidean domain.

3. Let F be a field and define $v(f(X))=\deg(f)$ for all nonzero $f(x) \in F[X]$. With this definition, $F[X]$ is a Euclidean domain.

You have probably noticed that we have not yet used Condition (i) in the definition of Euclidean domain. Condition (i) does have uses; you will find it helpful in Exercise 21.7, and it can also be used to give a direct proof that elements of a Euclidean domain can be factored into irreducibles, without using the fact that a Euclidean domain is a PID.

Observe that Theorems 21.5 and 21.6 can be summarized in the diagram

$$\text{Euclidean domain} \Rightarrow \text{PID} \Rightarrow \text{UFD}.$$

We have already seen that the second implication cannot be reversed, and neither can the first. The domain

$$D = \left\{ a + \frac{b}{2}(1 + \sqrt{-19}) \mid a, b \in \mathbb{Z} \right\}$$

is a PID, but it is impossible to define a function v from $D - \{0\}$ to $\mathbb{Z}^+ \cup \{0\}$ so that D becomes a Euclidean domain. [This example is due to T. Motzkin. See "The Euclidean Algorithm," *Bulletin of the American Mathematical Society*, vol. 55, pp. 1142–1146 (1949).]

We will illustrate the utility of the Euclidean domain concept by using it to prove that the ring $\mathbb{Z}[i]$ of Gaussian integers is a PID. Recall that

$$\mathbb{Z}[i] = \{a + bi \mid a, b \in \mathbb{Z}\}$$

(see Exercise 16.24).

Define $v(a+bi) = a^2 + b^2$, for $a + bi \neq 0$. By Exercise 16.24, v is multiplicative, that is, $v(\alpha\beta) = v(\alpha)v(\beta)$ for all nonzero $\alpha, \beta \in \mathbb{Z}[i]$. From this it follows that $v(\alpha) \leqslant v(\alpha\beta)$ for all nonzero α, β, because $v(\beta) \geqslant 1$. Thus Condition (i) in the definition of Euclidean domain is satisfied.

Now for Condition (ii). Let $\alpha, \beta \in \mathbb{Z}[i]$, $\beta \neq 0$. We seek to find γ and ρ in $\mathbb{Z}[i]$ such that $\alpha = \gamma\beta + \rho$, and either $\rho = 0$ or $v(\rho) < v(\beta)$. Since $\beta \neq 0$, α/β is a complex number, so there are real numbers x and y such that

$$\alpha = (x + yi)\beta.$$

Our idea is to use for γ an element of $\mathbb{Z}[i]$ that is close to $x + yi$. Specifically, choose integers a and b such that $|x - a| \leqslant \frac{1}{2}$ and $|y - b| \leqslant \frac{1}{2}$. Then

$$\alpha = (a + bi)\beta + [(x - a) + (y - b)i]\beta,$$

so if we take $\gamma = a + bi$ and $\rho = [(x - a) + (y - b)i]\beta$, we have

$$\alpha = \gamma\beta + \rho.$$

Now if $\rho = 0$, we are done. Otherwise $v[(x-a)+(y-b)i]$ is defined, and we have

$$v(\rho) = v[(x-a)+(y-b)i]v(\beta)$$
$$= [(x-a)^2+(y-b)^2]v(\beta)$$
$$\leqslant \left[\left(\frac{1}{2}\right)^2+\left(\frac{1}{2}\right)^2\right]v(\beta)$$
$$= \frac{1}{2}v(\beta) < v(\beta),$$

as desired.

Unique factorization in $\mathbb{Z}$ and $\mathbb{Z}[i]$ provide a very appealing proof of a classic theorem of Fermat. Fermat observed that some primes can be expressed in the form a^2+b^2 (with $a,b \in \mathbb{Z}$), while others cannot. For instance, $5 = 2^2+1^2$, $13 = 3^2+2^2$, $17 = 4^2+1^2$, $29 = 5^2+2^2$, $37 = 6^2+1^2$; but none of 3, 7, 11, 19, 23, or 31 can be written as the sum of two squares. It is also true that 5, 13, 17, 29, and 37 are all congruent to 1 (mod 4), while 3, 7, 11, 19, 23, and 31 are all congruent to 3. Fermat's Two Squares Theorem asserts that, in general, a positive odd prime p is the sum of two squares iff $p \equiv 1$ (mod 4).

Fermat claimed to have proved this result in a letter he wrote in 1640. In accordance with the practice of the time, he never published a proof, and the first published proof was given by Euler in 1754. It is said that Euler worked on and off for seven years to find a proof.

THEOREM 21.7 (Fermat) Let $p > 0$ be an odd prime. Then there exist integers a and b such that $p = a^2+b^2$ iff $p \equiv 1$ (mod 4). If $p \equiv 1$ (mod 4), then p can be written as a^2+b^2 in only one way (we do not count things like b^2+a^2 or $(-a)^2+(-b)^2$ as different ways.)

PROOF. Suppose $p = a^2+b^2$. Then $p \equiv a^2+b^2$ (mod 4), and each of a^2, b^2 is congruent (mod 4) to one of 0^2, 1^2, 2^2, or 3^2; that is, each of a^2, b^2 is $\equiv 0$ or 1 (mod 4). Since p is odd, one of a^2, b^2 is $\equiv 0$ and the other is $\equiv 1$ (mod 4). Hence $p \equiv 1$ (mod 4).

That was the easy half. To finish the proof, we must show that if $p \equiv 1$ (mod 4), then p can be written (uniquely) as a^2+b^2. The idea of our proof is to view a^2+b^2 as being $(a+bi)(a-bi)$ in $\mathbb{Z}[i]$, and to show that p can be written as $(a+bi)(a-bi)$ for some integers a,b.

Our first step is to show that p is at least not irreducible in $\mathbb{Z}[i]$. We know that if p *were* irreducible, then *since* $\mathbb{Z}[i]$ *is a PID*, p would be prime in $\mathbb{Z}[i]$. We are going to show that this is not the case by showing that there exists an integer m such that

$$(m-i)(m+i) = kp$$

for some integer k. Thus $p|(m-i)(m+i)$, but clearly p divides neither $(m-i)$ nor $(m+i)$ in $\mathbb{Z}[i]$, so p is not prime.

We find m by observing that having $(m-i)(m+i)=kp$ for some k is the same as having $m^2+1\equiv 0 \pmod{p}$, that is, $m^2\equiv -1 \pmod{p}$. We know *something* that is $\equiv -1 \pmod{p}$, namely $(p-1)!$ (see Wilson's Theorem, Exercise 10.26). Thus to find a suitable m it suffices to show that $(p-1)!\equiv m^2$ for some m. Now we know that $p=4n+1$ for some $n>0$, hence

$$(p-1)! = (1)(2)\cdots(2n)(2n+1)(2n+2)\cdots(4n-1)(4n)$$
$$\equiv (1)(2)\cdots(2n)(-2n)(-(2n-1))\cdots(-2)(-1)(\bmod\ p)$$
$$\equiv (-1)^{2n}1^2 2^2 3^2\cdots(2n)^2(\bmod\ p)$$
$$\equiv [(1)(2)(3)\cdots(2n)]^2(\bmod\ p).$$

Thus if we take $m=(1)(2)(3)\cdots(2n)$, we have the required m, and we see that p is not prime, hence not irreducible, in $\mathbb{Z}[i]$.

The rest is easy. Write

$$p=(a+bi)(c+di),$$

where neither $a+bi$ nor $c+di$ is a unit in $\mathbb{Z}[i]$. Then

$$v(p)=v(a+bi)v(c+di),$$

that is,

$$p^2=(a^2+b^2)(c^2+d^2).$$

Since neither $a+bi$ nor $c+di$ is a unit, neither a^2+b^2 nor c^2+d^2 is 1 (Exercise 16.24), so by unique factorization in $\mathbb{Z}$,

$$a^2+b^2=p=c^2+d^2,$$

and we have shown that p can be written as the sum of two squares. (Note that from $p=(a+bi)(c+di)$ it follows that $c=a$ and $d=-b$.)

All that remains is to establish the essential uniqueness of the representation. We leave this to you as Exercise 21.19.

EXERCISES

21.1 Let D be a domain. Show that it is not possible to express a unit of D as a product of prime elements.

21.2 Let D be a domain and let $a, b\in D$. Under what conditions is it true that $Da=Db$?

21.3 Let R be a UFD, $r\in R$. Show that r is irreducible in R iff it is irreducible in $R[X]$.

21.4 Complete the proof that $\mathbb{Z}[X]$ is a UFD by showing that every element which is neither 0 nor a unit can be written as a product of finitely many irreducible elements.

21.5 Show that every element $\neq 0$, ± 1 in $\mathbb{Z}[\sqrt{-3}\,]$ can be expressed as the product of finitely many irreducible elements.

21.6 Show that the following domains are not UFDs.

a) $D = \{a + b\sqrt{10} \mid a,b \in \mathbb{Z}\}$. (Consider $N(a + b\sqrt{10}) = a^2 - 10b^2$, and find two distinct factorizations of 6.)

b) $D = \{a + b\sqrt{-5} \mid a,b \in \mathbb{Z}\}$. (Consider $N(a + b\sqrt{-5}) = a^2 + 5b^2$, and find *three* distinct factorizations of 21.)

21.7 Let D be a Euclidean domain, with $v: D - \{0_D\} \to \mathbb{Z}^+ \cup \{0\}$.

a) Show that if $d \in D$, then d is a unit iff $v(d) = v(1_D)$.

b) Show that if v is a constant function then D is a field.

c) Show that, in general, if two nonzero elements a and b of a Euclidean domain are associates, then $v(a) = v(b)$.

21.8 Show that the following domains are Euclidean with the given function v.

a) $D = \{a + b\sqrt{2} \mid a,b \in \mathbb{Z}\}$, with $v(a + b\sqrt{2}) = |a^2 - 2b^2|$

b) $D = \{a + b\sqrt{-2} \mid a,b \in \mathbb{Z}\}$, with $v(a + b\sqrt{-2}) = a^2 + 2b^2$

21.9 Let D be a domain with the property that every element of D which is neither 0 nor a unit can be written as the product of a finite number of irreducibles. Show that D is a UFD iff every irreducible element of D is prime.

21.10 Let F be a field, and let X and Y be variables. Define the polynomial ring $F[X, Y]$ in two variables over F by

$$F[X, Y] = F[X][Y].$$

Thus $F[X, Y]$ is the ring of polynomials in Y, over $F[X]$.

a) Show that every element of $F[X, Y]$ is a finite sum of terms of the form aX^iY^j, where $a \in F$ and i,j are nonnegative integers.

b) Show that $F[X, Y]$ is not a PID.

21.11 Let D be a domain, and let $a,b \in D$. An element $d \in D$ is called a *greatest common divisor* (g.c.d.) of a and b if

i) $d \mid a$ and $d \mid b$; and

ii) if c is any element such that $c \mid a$ and $c \mid b$, then $c \mid d$.

Show that if D is a PID, then any two elements of D have a g.c.d. which can be written in the form $xa + yb$, for some $x,y \in D$. (*Suggestion:* Consider the ideal $\{xa + yb \mid x,y \in D\}$.)

21.12 In the context of $\mathbb{Z}$, the definition in Exercise 21.11 assigns two g.c.d.'s to each pair of nonzero elements a,b. For instance, the g.c.d.'s of 12 and 15 are 3 and -3. (Compare Exercise 4.26.) Show that, in general, if D is a domain, and $d \in D$ is a g.c.d. of a and b, then all the g.c.d.'s of a and b are precisely the associates of d.

21.13 Show that the elements 4 and $2(1 + \sqrt{-3})$ in $\mathbb{Z}[\sqrt{-3}]$ have no g.c.d.

21.14 Use the Euclidean algorithm (that is, repeated application of the division algorithm) in $\mathbb{Q}[X]$ to find a g.c.d. of the elements $2X^3 + 9X^2 + 12X + 5$ and $2X^5 + 5X^4 + 8X + 20$.

21.15 Let $\alpha \in \mathbb{Z}[i]$. Show that α is a prime element in $\mathbb{Z}[i]$ iff either $v(\alpha)$ is a prime integer or α is an associate of some prime integer p such that $p \equiv 3 \pmod 4$.

21.16 Write each of the following elements of $\mathbb{Z}[i]$ as a product of primes (see Exercise 21.15).

a) $1 + 3i$

b) $7 + 8i$

c) $99 + 27i$

21.17 a) Prove that if $a, b \in \mathbb{Z}$ and p is a prime integer, $p \equiv 3 \pmod 4$, such that $p | (a^2 + b^2)$, then $p^2 | (a^2 + b^2)$.
   (Use the fact that p is prime in $\mathbb{Z}[i]$.)

b) Prove that an integer $n \geqslant 2$ is the sum of two (integer) squares iff in the prime factorization of n (in $\mathbb{Z}$), every prime $p \equiv 3 \pmod 4$ occurs to an even power.

21.18 a) Use the Euclidean algorithm in $\mathbb{Z}[i]$ to find a g.c.d. for $53 + 9i$ and $1 + 7i$.
   (See the proof that $\mathbb{Z}[i]$ is a Euclidean domain, and use the fact that for complex numbers $a + bi$ and $c + di$,

$$\frac{a + bi}{c + di} = \frac{(a + bi)(c - di)}{c^2 + d^2}.\Big)$$

b) The proof that $\mathbb{Z}[i]$ is Euclidean provides an upper bound on the number of steps it will take the Euclidean algorithm to produce a g.c.d. What is this bound?

21.19 Prove the uniqueness statement of Theorem 21.7.

21.20 Give an example of two domains D and D' such that D is a UFD, D' is a homomorphic image of D, and D' is not a UFD.

21.21 Let R be a PID and I an ideal of R.

a) Show that every ideal of R/I is principal. Must R/I be a PID?

b) Show that R/I has only finitely many ideals if I is nontrivial.

21.22 Show that $\sqrt{-3}$ is prime in $\mathbb{Z}[\sqrt{-3}]$.

21.23 Let

$$\zeta_3 = \cos(2\pi/3) + i\sin(2\pi/3) = -\tfrac{1}{2} + i(\sqrt{3})/2 = \frac{-1 + \sqrt{-3}}{2}.$$

Let

$$\mathbb{Z}[\zeta_3] = \{a + b\zeta_3 + c\zeta_3^2 | a, b, c \in \mathbb{Z}\},$$

and observe that $\zeta_3^3 = 1$. Thus $\mathbb{Z}[\zeta_3]$ is the domain considered by Kummer in his work on Fermat's Last Theorem, for $p = 3$. This exercise will result in a proof that $\mathbb{Z}[\zeta_3]$ is a Euclidean domain. [A proof of Fermat's Last Theorem for $p = 3$, using $\mathbb{Z}[\zeta_3]$, can be found in Hardy and Wright, *An Introduction to the Theory of Numbers*, Chapter XIII (Oxford: Clarendon Press, 1960).]

a) Show that $\mathbb{Z}[\zeta_3] = \{a + b\zeta_3 | a, b \in \mathbb{Z}\}$, and that every element of $\mathbb{Z}[\zeta_3]$ has a unique representation in the form $a + b\zeta_3$.

    b) Let us write just ζ instead of ζ_3, for simplicity. Define

$$v(a+b\zeta)=(a+b\zeta)(a+b\zeta^2).$$

    Show that $v(a+b\zeta)=a^2+b^2-ab$, and that v is multiplicative, that is, $v(\alpha\beta)=v(\alpha)v(\beta)$.

    c) Show that v maps $\mathbb{Z}[\zeta]-\{0\}$ into $\mathbb{Z}^+$, and that $\mathbb{Z}[\zeta]$, with the function v, is a Euclidean domain.

21.24 a) Show that $\mathbb{Z}[\sqrt{-3}\,]\subseteq\mathbb{Z}[\zeta_3]$. (See the preceding exercise.)

    b) Determine the relationship between $\mathbb{Z}[\sqrt{-3}\,]$ and $\mathbb{Z}[\zeta_3]$ by describing the elements of $\mathbb{Z}[\zeta_3]-\mathbb{Z}[\sqrt{-3}\,]$ in terms of integers and $\sqrt{-3}$.

21.25 See Exercise 21.23.

    a) Show that if $\alpha\in\mathbb{Z}[\zeta_3]$, then α is a unit iff $v(\alpha)=1$.

    b) Show that if $\alpha\in\mathbb{Z}[\zeta_3]$ and $v(\alpha)$ is a prime integer, then α is irreducible. Conclude that $1-\zeta_3$ is irreducible.

    c) Show that 2 is prime in $\mathbb{Z}[\zeta_3]$. (*Hint:* Show that if $2|(a^2+b^2-ab)$, then both a and b are even.) There are many more primes in $\mathbb{Z}[\zeta_3]$; see Hardy and Wright, Chapter XV.

21.26 Let D be a Euclidean domain. Show that the q and r in Condition (ii) for a Euclidean domain will be unique for every choice of a and b iff v has the property that $v(a+b)\leqslant\max\{v(a),v(b)\}$ for all nonzero a and b such that $a+b\neq0$.

SUGGESTIONS FOR FURTHER READING

Algebra

Artin, E. (1944). *Galois Theory*, 2d ed. Notre Dame Mathematical Lectures Number 2. Notre Dame, Indiana: University of Notre Dame Press.

Burton, D. M. (1970). *A First Course in Rings and Ideals*. Reading, Mass.: Addison-Wesley.

Fraleigh, J. B. (1976). *A First Course in Abstract Algebra*, 2d ed. Reading, Mass.: Addison-Wesley.

Gaal, L. (1971). *Classical Galois Theory with Examples*. Chicago: Markham. (Reprinted by Chelsea, New York, 1973.)

Goldstein, L. J. (1973). *Abstract Algebra: A First Course*. Englewood Cliffs, N.J.: Prentice-Hall.

Hadlock, C. R. (1978). *Field Theory and its Classical Problems*. Carus Mathematical Monographs, No. 19. Providence, R.I.: The Mathematical Association of America.

Hall, M., Jr. (1959). *The Theory of Groups*. New York: Macmillan. (Reprinted by Chelsea, New York, 1976.)

Herstein, I. N. (1975). *Topics in Algebra*, 2d ed. Lexington, Mass.: Xerox College Publishing.

Jacobson, N. (1974). *Basic Algebra I*. San Francisco: W. H. Freeman.

Lang, S. (1965). *Algebra*. Reading, Mass.: Addison-Wesley.

Rotman, J. (1973). *The Theory of Groups: An Introduction*, 2d ed. Boston: Allyn and Bacon.

Schilling, O. F. G., and W. S. Piper (1975). *Basic Abstract Algebra*. Boston: Allyn and Bacon.

Zariski, O., and P. Samuel (1975). *Commutative Algebra I*. New York: Springer-Verlag.

Number Theory

Burton, D. M. (1976). *Elementary Number Theory*. Boston: Allyn and Bacon.

Dudley, U. (1978). *Elementary Number Theory*, 2d ed. San Francisco: W. H. Freeman.

Edwards, H. M. (1977). *Fermat's Last Theorem*. New York: Springer-Verlag.

Hardy, G. H., and E. M. Wright (1960). *An Introduction to the Theory of Numbers*, 4th ed. Oxford: Clarendon Press.

Pollard, H., and H. G. Diamond (1975). *The Theory of Algebraic Numbers*, 2d ed. Carus Mathematical Monographs, No. 9. Providence, R.I.: The Mathematical Association of America.

Sierpinski, W. (1964). *Elementary Theory of Numbers*. New York: Hafner.

History

Bell, E. T. (1945). *The Development of Mathematics*, 2d ed. New York: McGraw-Hill.

Bell, E. T. (1965). *Men of Mathematics*. New York: Simon and Schuster.

Boyer, C. A. (1968). *A History of Mathematics*. New York: Wiley.

Cajori, F. (1919). *A History of Mathematics*. New York: Macmillan.

Dickson, L. E. (1919, 1920, 1923). *History of the Theory of Numbers*, 3 vols. Washington: Carnegie Institute. (Reprinted by Chelsea, New York, 1971.)

Eves, H. W. (1969). *An Introduction to the History of Mathematics*, 3d ed. New York: Holt, Rinehart, and Winston.

ANSWERS TO
SELECTED EXERCISES

Section 1

1.1 (a) $\{2, 5, \pi, 5/2, 4, 6, 3/2\}$

(b) $\left\{ \begin{pmatrix} 1 & 3 \\ 4 & 6 \end{pmatrix}, \begin{pmatrix} 5 & 8 \\ 0, & -1 \end{pmatrix}, \begin{pmatrix} 1 & 1 \\ 1 & \pi \end{pmatrix}, \begin{pmatrix} 5 & 8 \\ 0 & 1 \end{pmatrix}, \begin{pmatrix} 1 & 2 \\ 3 & 4 \end{pmatrix} \right\}$

1.3 (a) Yes;   (d) no;   (e) yes;   (h) no;   (j) yes

1.4 No; no

1.5 Yes; yes

1.6 (a) Not commutative; not associative

(e) Commutative; associative

(j) Commutative only if $X = \varnothing$; associative

1.9 Yes; no

Section 2

2.1 (b), (d), (e), (g), (h), and (i) are groups

2.2 (a) All of them;   (b) all except $GL(2, \mathbb{R})$

2.3 Only if $X = \varnothing$

2.4 (a)

$\oplus$	0	1	2	3
0	0	1	2	3
1	1	2	3	0
2	2	3	0	1
3	3	0	1	2

2.5 No

2.8 It is a group.

Section 3

3.1 $x = 4$

3.2 $x = \{1, 2, 5, 6, 7, 8\}$

Section 4

4.1 $\langle 3 \rangle = \mathbb{Z}_{10}$; $\langle 8 \rangle = \{8,6,4,2,0\}$

4.3 $\langle A \rangle = \{A, \varnothing\}$

4.4   10, 15, 5, 30, and 5, respectively

4.5   9, 6, 9, 18, and 3, respectively

4.6   3, 6, 12, 21, 24, 33, 39, 42

4.7 x^6, x^{18}

4.8   Yes

4.10   (b) Yes

4.15   (b) $(862, 347) = 1$ and we can use $x = 126$, $y = -313$

Section 5

5.1   (a), (b), (d), (f), (h), and (i) are subgroups

5.4   (a) Six. They are $\langle 1 \rangle$, $\langle 2 \rangle$, $\langle 3 \rangle$, $\langle 6 \rangle$, $\langle 9 \rangle$, and $\langle 0 \rangle$.

5.8   72

5.9 $n|m$; one of m,n divides the other

5.12   (b) $\{I, -I\}$

5.13 $\left\{\begin{pmatrix} a & 0 \\ 0 & a \end{pmatrix} \middle| a \neq 0\right\}$

Section 6

6.1   (a) 18;   (b) 24;   (c) 18;   (d) 765

6.2   The group in (d) is cyclic.

6.11   You should end up with 16 distinct subgroups.

Section 7

7.1   (a) Function from S to T; neither one-to-one nor onto

(b) Not a function from S to T

(c) Not a function

(d) Function from S to T; neither one-to-one nor onto

(e) Function from S to T; one-to-one and onto

(f) Not a function from S to T

(g) Not a function from S to T

(h) Function from S to T; one-to-one but not onto

(i) Function from S to T; onto but not one-to-one

(j) Function from S to T; neither one-to-one nor onto

7.4   Onto, not one-to-one

7.5 $A = X$

7.6   Yes

7.8   Yes, on all counts

Section 8

8.1 (b) $\begin{pmatrix} 1 & 2 & 3 & 4 & 5 & 6 \\ 6 & 1 & 4 & 3 & 2 & 5 \end{pmatrix}$

8.2 (a) $\begin{pmatrix} 1 & 2 & 3 & 4 & 5 & 6 \\ 3 & 6 & 1 & 4 & 2 & 5 \end{pmatrix} = (1,3)(2,6,5) = (1,3)(2,5)(2,6)$. Odd.

(c) $\begin{pmatrix} 1 & 2 & 3 & 4 & 5 & 6 \\ 5 & 6 & 3 & 4 & 1 & 2 \end{pmatrix} = (1,5)(2,6)$. Even.

8.3 (a) Odd r's

(b) Factor it into cycles. The permutation is even iff the number of r-cycles with even r's is even.

(c) Even

8.10 (c) 12   **8.11** (a) $\{e\}$;   (b) $\{e, f^2\}$ **8.12** (b) f^3;   (c) $\{e\}$

Section 9

9.1 The relation in (b) is an equivalence relation.

9.4 (a) $H = \{J, -I, -J, I\}$ and $H \cdot K = \{L, -K, -L, K\}$

9.5 The right cosets are $H = \{e, f^2g\}$, $Hf = \{f, fg\} = Hfg$, $Hf^2 = \{f^2, g\} = Hg$, and $Hf^3 = \{f^3, f^3g\}$.

9.6 $H = \{(0,0),(1,0),(2,0)\}$ and $H + (0,1) = \{(0,1),(1,1),(2,1)\} = H + (1,1) = H + (2,1)$.

Section 10

10.1 4, 2, and 2   **10.2** (a) 16;   (b) 6;   (c) 4

10.3 (a) 6;   (b) 4   **10.4** 4

10.21 The conjugacy classes are $\{I\}$, $\{-I\}$, $\{J, -J\}$, $\{K, -K\}$, and $\{L, -L\}$. The class equation is $8 = 2 + 2 + 2 + 2$.

Section 11

11.2 No

11.10 6

11.13 It is essentially the same as V.

11.18 Just a remark: You are given more information than you need to answer the question.

Section 12

12.1 (a) Epimorphism;   (b) isomorphism;   (c) epimorphism;   (d) epimorphism; (e) not a homomorphism

12.3 No; no

12.4 (a) No;   (b) yes;   (c) no;   (e) no;   (g) yes;   (h) yes

12.8 No; yes

12.9 No

12.17 6

Section 13

13.1 $\ker(\varphi) = \{0, 4\}$. The quotient group is isomorphic to $(\mathbb{Z}_4, \oplus)$.

13.3 D_4, V, $(\mathbb{Z}_2, \oplus)$, and the trivial group

13.6 G has 3 elements.

Section 14

14.1 (b) $\mathbb{Z}_8 \times \mathbb{Z}_9$
$\mathbb{Z}_8 \times \mathbb{Z}_3 \times \mathbb{Z}_3$
$\mathbb{Z}_4 \times \mathbb{Z}_2 \times \mathbb{Z}_9$
$\mathbb{Z}_4 \times \mathbb{Z}_2 \times \mathbb{Z}_3 \times \mathbb{Z}_3$
$\mathbb{Z}_2 \times \mathbb{Z}_2 \times \mathbb{Z}_2 \times \mathbb{Z}_9$
$\mathbb{Z}_2 \times \mathbb{Z}_2 \times \mathbb{Z}_2 \times \mathbb{Z}_3 \times \mathbb{Z}_3$

(d) $\mathbb{Z}_2 \times \mathbb{Z}_9 \times \mathbb{Z}_{25}$
$\mathbb{Z}_2 \times \mathbb{Z}_9 \times \mathbb{Z}_5 \times \mathbb{Z}_5$
$\mathbb{Z}_2 \times \mathbb{Z}_3 \times \mathbb{Z}_3 \times \mathbb{Z}_{25}$
$\mathbb{Z}_2 \times \mathbb{Z}_3 \times \mathbb{Z}_3 \times \mathbb{Z}_5 \times \mathbb{Z}_5$

Section 15

15.8 $\mathbb{Z}_{25} \times \mathbb{Z}_{169}$
$\mathbb{Z}_{25} \times \mathbb{Z}_{13} \times \mathbb{Z}_{13}$
$\mathbb{Z}_5 \times \mathbb{Z}_5 \times \mathbb{Z}_{169}$
$\mathbb{Z}_5 \times \mathbb{Z}_5 \times \mathbb{Z}_{13} \times \mathbb{Z}_{13}$

Section 16

16.2 (a) No; (b) no

16.4 Yes

16.11 (b) Units: $(1, 1), (1, 2), (2, 1), (2, 2)$;
zero-divisors: $(0, 0), (0, 1), (0, 2), (1, 0), (2, 0)$;
nilpotent elements: $(0, 0)$

16.20 (a) No

Section 17

17.1 The sets in (b) and (d) are subrings.

17.2 (a) Ideal; (b) not a subring; (c) subring, not an ideal

17.3 The subgroups of G

17.6 (a) $\{0, 3\}$, the principal ideal generated by 3; or $\{0, 2, 4\}$, the principal ideal generated by 2 or 4.

Section 18

18.1 The mapping in (b) is a ring homomorphism.

18.2 No; yes

18.7 Up to isomorphism, they are $\{\mathbb{Z}_d \mid d$ a positive divisor of $n\}$.

Section 19

19.2 (a) Irreducible

(e) Irreducible

19.3 (a) Irreducible

(d) $(X+1)(X^3+2X+2)$

Section 20

20.2 No

20.5 $(ac+bd)+(ad+bc+bd)\bar{X}$

20.7 (a) Let $F=\mathbb{Z}_2$, let $f(X)=X^3+X+1$, and let K be the field described in Exercise 20.6.

Section 21

21.14 $2X+5$

21.16 (a) $(-1+i)(1-2i)$

21.18 (a) $-1+i$

INDEX

LIST OF SPECIAL SYMBOLS

$\mathbb{Z}$	set of all integers
$\mathbb{Q}$	set of all rational numbers
$\mathbb{R}$	set of all real numbers
$\mathbb{C}$	set of all complex numbers
$\Leftrightarrow$, iff	if and only if
$\mathbb{Z}^+$, $\mathbb{Q}^+$, $\mathbb{R}^+$	sets of all positive integers, rational numbers, and real numbers, respectively
$a\|b$	a divides b
$*$	binary operation
$\triangle$	symmetric difference
e	identity element of a group
$GL(2, \mathbb{R})$	general linear group of degree 2 over $\mathbb{R}$
$P(X)$	set of subsets X
$\mathbb{Z}_n$	the set $\{0, 1, 2, \ldots, n-1\}$
$a \equiv b \pmod{n}$	the integers a and b are congruent modulo n
$\oplus$, $\odot$	addition and multiplication modulo n
$o(x)$	order of the element x
$<x>$	set of powers of the element x
$\|G\|$	order of the group G
V	Klein's 4-group
$Z(G)$	center of the group G
$GL(2, \mathbb{C})$	general linear group of degree 2 over $\mathbb{C}$
Q_8	group of unit quaternions
$SL(2, \mathbb{R})$	special linear group of degree 2 over $\mathbb{R}$
$Z(g)$	centralizer of the element g
$G \times H$	direct product of G and H